ENVIRONMENTAL RESEARCH ADVANCES

AF587695

PRESIDENT OBAMA'S CLIMATE ACTION PLAN

ELEMENTS AND ANALYSES

ENVIRONMENTAL RESEARCH ADVANCES

Additional books in this series can be found on Nova's website under the Series tab.

Additional E-books in this series can be found on Nova's website under the E-book tab.

CLIMATE CHANGE AND ITS CAUSES, EFFECTS AND PREDICTION

Additional books in this series can be found on Nova's website under the Series tab.

Additional E-books in this series can be found on Nova's website under the E-book tab.

ENVIRONMENTAL RESEARCH ADVANCES

PRESIDENT OBAMA'S CLIMATE ACTION PLAN

ELEMENTS AND ANALYSES

SHANE N. BLAKE
EDITOR

New York

Copyright © 2013 by Nova Science Publishers, Inc.

All rights reserved. No part of this book may be reproduced, stored in a retrieval system or transmitted in any form or by any means: electronic, electrostatic, magnetic, tape, mechanical photocopying, recording or otherwise without the written permission of the Publisher.

For permission to use material from this book please contact us:
Telephone 631-231-7269; Fax 631-231-8175
Web Site: http://www.novapublishers.com

NOTICE TO THE READER

The Publisher has taken reasonable care in the preparation of this book, but makes no expressed or implied warranty of any kind and assumes no responsibility for any errors or omissions. No liability is assumed for incidental or consequential damages in connection with or arising out of information contained in this book. The Publisher shall not be liable for any special, consequential, or exemplary damages resulting, in whole or in part, from the readers' use of, or reliance upon, this material. Any parts of this book based on government reports are so indicated and copyright is claimed for those parts to the extent applicable to compilations of such works.

Independent verification should be sought for any data, advice or recommendations contained in this book. In addition, no responsibility is assumed by the publisher for any injury and/or damage to persons or property arising from any methods, products, instructions, ideas or otherwise contained in this publication.

This publication is designed to provide accurate and authoritative information with regard to the subject matter covered herein. It is sold with the clear understanding that the Publisher is not engaged in rendering legal or any other professional services. If legal or any other expert assistance is required, the services of a competent person should be sought. FROM A DECLARATION OF PARTICIPANTS JOINTLY ADOPTED BY A COMMITTEE OF THE AMERICAN BAR ASSOCIATION AND A COMMITTEE OF PUBLISHERS.

Additional color graphics may be available in the e-book version of this book.

Library of Congress Cataloging-in-Publication Data

ISBN: 978-1-62808-863-2

Published by Nova Science Publishers, Inc. † New York

CONTENTS

PREFACE

On June 25, 2013, President Obama announced a national plan to reduce emissions of carbon dioxide (CO2) and other greenhouse gases (GHG), as well as to encourage adaptation to expected climate change. The President affirmed his commitment to his 2009 policy pledge to reduce U.S. GHG emissions by 17% below 2005 levels by 2020 if all other major economies agreed to limit their emissions as well. In 2011, the United States' gross GHG emissions were approximately 7% below their 2005 levels. The President stated a willingness to work with Congress toward enacting a bipartisan, market-based scheme to reduce GHG emissions. The President's Climate Action Plan lays out a series of measures in three categories: 1) cut carbon pollution in America, 2) prepare the United States for the impacts of climate change and 3) lead international efforts to address global climate change. Many measures included in the Climate Action Plan have been underway. The plan specifies few timelines or metrics for evaluating progress of individual measures beyond national aggregate or sectoral GHG emissions or energy efficiency. The centerpiece of the President's announcement arguably is a Presidential Memorandum, also issued June 25, that directs EPA to issue two types of rules to curtail carbon dioxide emissions from new and existing power plants before the end of his term. This book provides an overview of President Obama's Climate Action Plan, with a focus on key elements and analyses.

Chapter 1 – On June 25, 2013, President Obama announced a national plan to reduce emissions of carbon dioxide (CO2) and other greenhouse gases (GHG), as well as to encourage adaptation to expected climate change. The President affirmed his commitment to his 2009 policy pledge to reduce U.S. GHG emissions by 17% below 2005 levels by 2020 if all other major economies agreed to limit their emissions as well. In 2011, the United States' gross GHG emissions were approximately 7% below their 2005 levels. The President stated a willingness to work with Congress toward enacting a bipartisan, market-based scheme to reduce GHG emissions. He also said that he would move ahead with Executive Branch actions in the absence of Congressional support. The President's Climate Action Plan lays out a series of measures in three categories: Cut carbon pollution in America. Prepare the United States for the impacts of climate change. Lead international efforts to address global climate change. Many measures included in the Climate Action Plan have been underway. The plan specifies few timelines or metrics for evaluating progress of individual measures beyond national aggregate or sectoral GHG emissions or energy efficiency. The centerpiece of the President's announcement arguably is a Presidential Memorandum, also issued June 25, that directs EPA to issue two types of rules to curtail carbon dioxide emissions from new and existing power plants before the end of his term. Specifically, the Presidential Memorandum first instructs EPA to issue, as planned, a new proposal under the Clean Air Act (CAA), by September 20, 2013, for GHG emissions from newly constructed electric generating units (EGU), and to issue the final rule "in a timely fashion" after comments. Second, and most significantly, the Presidential Memorandum directs EPA also to issue standards, regulations, or guidelines for CO2 emissions applicable to modified, reconstructed and existing power plants, building on States' efforts to reduce power plant emissions. The Memorandum included neither specific levels of EGU emissions performance nor target-sectoral GHG reductions. The Memorandum requests the proposed rules for existing EGU by June 1, 2014, and final rules by June 1, 2015. Further, the Memorandum requests that the guidelines require States to submit to EPA their implementation plans, required under §111(d) of the CAA, and their implementing regulations by June 30, 2016. The language of the announcement and Memorandum suggest that the new standards and guidelines might be written to allow innovative, potentially cost-cutting flexibilities to states and regulated entities. The President's Climate Action Plan additionally announced regulatory actions to: reduce GHG emissions including fuel efficiency standards for heavy-duty vehicles post-2018; tighten efficiency standards for federal buildings; and require a transition away from chemicals that contribute to global climate change that were introduced as alternatives to stratospheric ozone-depleting chemicals. In President Obama's speech, he referred to the pending Presidential Permit to allow construction of the Keystone XL pipeline to carry Canadian oil sands across the U.S. border. President Obama stated, "The net effects of the pipeline's impact on our climate will be absolutely critical to determining whether this project is allowed to go forward." A host of administrative actions would promote GHG emission reduction, energy efficiency, and increased electricity generation by renewable energy in federal facilities, on federal lands, and among private, state, and local partners of federal agencies. Budget proposals for some of these actions were included in the President's proposal for FY2014. To promote adaptation to climate change by the federal government and states and localities, the Climate Action Plan includes mostly a continuation of existing programs. Additionally, a set of existing international initiatives were included in the President's announcement to promote global reductions of GHG emissions and adaptation. New in the President's announcement is a call to end U.S. support for public financing of new coal-fired power plants overseas except for those employing advanced efficiency or carbon capture and sequestration technology. Notably, the plan provided no quantification of whether the United States would meet its commitment to reduce GHG emissions by 17% from 2005 levels by 2020 or whether or how the United States would work towards its share of a 2009 international pledge of $100 billion annually to assist developing countries to mitigate their GHG emissions and adapt to climate change.

Chapter 2 – Repors on Executive Office of the President publication, on The President's Climate Action Plan, dated June 2013.

Chapter 3 – Report on The White House, Office of the Press Secretary, on Presidential Memorandum - Power Sector Carbon Pollution Standards, dated June 25, 2013.

Chapter 4 – As President Obama announced initiatives addressing climate change on June 25, 2013, a major focus of attention was on the prospect of greenhouse gas (GHG) emission standards for fossilfueled—mostly coal-fired—electric generating units (EGUs). EGUs (more commonly referred to as power plants) are the largest source of greenhouse gas emissions, accounting for about one-third of total U.S. GHGs. If the country is going to reduce its GHG emissions by significant amounts, as the President has committed to do, emissions from these sources will almost certainly need to be controlled. The President addressed this issue on June 25 by directing EPA to re-propose GHG emission standards for new EGUs, which the agency had proposed in April 2012, but had not yet finalized. He also directed the agency to develop standards for existing power plants by June 2015. Under the Clean Air Act, the EPA Administrator has a great deal of flexibility in setting these standards. The statute requires that New Source Performance Standards (NSPS) reflect the degree of emission limitation achievable through application of the best system of emission reduction that has been "adequately demonstrated." The Administrator can take costs, health and environmental impacts, and energy requirements into account in determining what has been adequately demonstrated. The standard for new EGUs proposed in 2012 would have set a limit of 1,000 pounds of carbon dioxide (CO2) per megawatt-hour of electricity generated—a standard that can be met by new natural gas combined cycle plants without add-on emission controls. Coal-fired plants, however, would find it impossible to meet the standard without controls to capture, compress, and store underground about 45% of the CO2 they produce—a technology referred to as carbon capture and storage (CCS). Many in the electric power and coal industries view the 2012 proposed standard as effectively prohibiting the construction of new coal-fired power plants. Whether CCS technology has been adequately demonstrated is one question they raise. Other issues involve the cost of compliance, the increased energy required to capture and store carbon, and whether the agency should propose separate standards for gas-fired and coal-fired units. These questions may be addressed in the re-proposal, which the President directed to be issued no later than September 20, 2013. In its 2012 proposal, EPA maintained that the components of CCS technology have been demonstrated on numerous facilities. Despite this, the agency concluded that no new facilities (other than DOE-sponsored demonstration projects) will actually use CCS in the next 10 years. Given the projected low cost and abundance of natural gas, all new fossil-fueled units are likely to be powered by gas, according to EPA. Interest in the power plant proposal extends far beyond the electric power and energy industries. Power plants are the first category of sources for which EPA has proposed to require CCS, but the agency expects to propose standards for the GHG emissions of other industries in the next few years. What the agency promulgates in this instance may serve as a precedent. The potential impacts of the rule also extend beyond new sources, because the agency is obligated under Section 111(d) of the Clean Air Act to promulgate guidelines for existing sources within a category when it promulgates GHG standards for new sources. Using these guidelines, states will be required to develop performance standards for existing sources. These could be less stringent than the NSPS—taking into account, among other factors, the remaining useful life of the existing source—but the standards could have far greater impact than the NSPS, given that they will affect all existing sources. Many in Congress oppose GHG emission standards. In the 112th Congress, the House passed bills (H.R. 1, H.R. 910, and H.R. 3409) that would have prohibited EPA from promulgating GHG emission standards for any source; the Senate did not follow suit. The President's climate change initiative is likely to stir renewed interest in this subject.

Chapter 5 – Carbon capture and sequestration (or storage)—known as CCS—has attracted congressional interest as a measure for mitigating global climate change because large amounts of carbon dioxide (CO_2) emitted from fossil fuel use in the United States are potentially available to be captured and stored underground and prevented from reaching the atmosphere. Large, industrial sources of CO_2, such as electricity-generating plants, are likely initial candidates for CCS because they are predominantly stationary, single-point sources. Electricity generation contributes over 40% of U.S. CO_2 emissions from fossil fuels. Currently, U.S. power plants do not capture large volumes of CO_2 for CCS. Several projects in the United States and abroad—typically associated with oil and gas production—are successfully capturing, injecting, and storing CO_2 underground, albeit at relatively small scales. The oil and gas industry in the United States injects nearly 50 million tons of CO_2 underground each year for the purpose of enhanced oil recovery (EOR). The volume of CO_2 envisioned for CCS as a climate mitigation option is overwhelming compared to the amount of CO_2 used for EOR. According to the U.S. Department of Energy (DOE), the United States has the potential to store billions of tons of CO_2 underground and keep the gas trapped there indefinitely. Capturing and storing the equivalent of decades or even centuries of CO_2 emissions from power plants (at current levels of emissions) suggests that CCS has the potential to reduce U.S. greenhouse gas emissions substantially while allowing the continued use of fossil fuels. An integrated CCS system would include three main steps: (1) capturing and separating CO_2 from other gases; (2) purifying, compressing, and transporting the captured CO_2 to the sequestration site; and (3) injecting the CO_2 in subsurface geological reservoirs or storing it in the oceans. Deploying CCS technology on a commercial scale would be a vast undertaking. The CCS process, although simple in concept, would require significant investments of capital and of time. Capital investment would be required for the technology to capture CO_2 and for the pipeline network to transport the captured CO_2 to the disposal site. Time would be required to assess the potential CO_2 storage reservoir, inject the captured CO_2, and monitor the injected plume to ensure against leaks to the atmosphere or to underground sources of drinking water, potentially for years or decades until injection activities cease and the injected plume stabilizes. Three main types of geological formations in the United States are being considered for storing large amounts of CO_2: oil and gas reservoirs, deep saline reservoirs, and unmineable coal seams. The deep ocean also has a huge potential to store carbon; however, direct injection of CO_2 into the deep ocean is controversial, and environmental concerns have forestalled planned experiments in the open ocean. Mineral carbonation—reacting minerals with a stream of concentrated CO_2 to form a solid carbonate—is well understood, but it is still an experimental process for storing large quantities of CO_2. Large-scale CCS injection experiments are only beginning in the United States to test how different types of reservoirs perform during CO_2 injection of a million tons of CO_2 or more. Results from the experiments will undoubtedly be crucial to future permitting and site approval regulations. Acceptance by the general public of large-scale deployment of CCS may be a significant challenge. Some of the large-scale injection tests could garner information about public acceptance, as citizens become familiar with the concept, process, and results of CO_2 injection tests in their local communities.

In: President Obama's Climate Action Plan
Editor: Shane N. Blake
ISBN: 978-1-62808-863-2
© 2013 Nova Science Publishers, Inc.

Chapter 1

PRESIDENT OBAMA'S CLIMATE ACTION PLAN[*]

Jane A. Leggett

SUMMARY

On June 25, 2013, President Obama announced a national plan to reduce emissions of carbon dioxide (CO2) and other greenhouse gases (GHG), as well as to encourage adaptation to expected climate change. The President affirmed his commitment to his 2009 policy pledge to reduce U.S. GHG emissions by 17% below 2005 levels by 2020 if all other major economies agreed to limit their emissions as well. In 2011, the United States' gross GHG emissions were approximately 7% below their 2005 levels.

The President stated a willingness to work with Congress toward enacting a bipartisan, market-based scheme to reduce GHG emissions. He also said that he would move ahead with Executive Branch actions in the absence of Congressional support. The President's Climate Action Plan lays out a series of measures in three categories:

- Cut carbon pollution in America.
- Prepare the United States for the impacts of climate change.
- Lead international efforts to address global climate change.

---

[*] This is an edited, reformatted and augmented version of a Congressional Research Service publication, CRS Report for Congress R43120, prepared for Members and Committees of Congress, from www.crs.gov, dated June 26, 2013.

Many measures included in the Climate Action Plan have been underway. The plan specifies few timelines or metrics for evaluating progress of individual measures beyond national aggregate or sectoral GHG emissions or energy efficiency.

The centerpiece of the President's announcement arguably is a Presidential Memorandum, also issued June 25, that directs EPA to issue two types of rules to curtail carbon dioxide emissions from new and existing power plants before the end of his term. Specifically, the Presidential Memorandum first instructs EPA to issue, as planned, a new proposal under the Clean Air Act (CAA), by September 20, 2013, for GHG emissions from newly constructed electric generating units (EGU), and to issue the final rule "in a timely fashion" after comments. Second, and most significantly, the Presidential Memorandum directs EPA also to issue standards, regulations, or guidelines for CO2 emissions applicable to modified, reconstructed and existing power plants, building on States' efforts to reduce power plant emissions. The Memorandum included neither specific levels of EGU emissions performance nor target-sectoral GHG reductions. The Memorandum requests the proposed rules for existing EGU by June 1, 2014, and final rules by June 1, 2015. Further, the Memorandum requests that the guidelines require States to submit to EPA their implementation plans, required under §111(d) of the CAA, and their implementing regulations by June 30, 2016. The language of the announcement and Memorandum suggest that the new standards and guidelines might be written to allow innovative, potentially cost-cutting flexibilities to states and regulated entities.

The President's Climate Action Plan additionally announced regulatory actions to:

- reduce GHG emissions including fuel efficiency standards for heavy-duty vehicles post-2018;
- tighten efficiency standards for federal buildings; and require a transition away from chemicals that contribute to global climate change that were introduced as alternatives to stratospheric ozone-depleting chemicals.

In President Obama's speech, he referred to the pending Presidential Permit to allow construction of the Keystone XL pipeline to carry Canadian oil sands across the U.S. border. President Obama stated, "The net effects of the pipeline's impact on our climate will be absolutely critical to determining whether this project is allowed to go forward."

A host of administrative actions would promote GHG emission reduction, energy efficiency, and increased electricity generation by renewable energy in federal facilities, on federal lands, and among private, state, and local partners of federal agencies. Budget proposals for some of these actions were included in the President's proposal for

FY2014. To promote adaptation to climate change by the federal government and states and localities, the Climate Action Plan includes mostly a continuation of existing programs.

Additionally, a set of existing international initiatives were included in the President's announcement to promote global reductions of GHG emissions and adaptation. New in the President's announcement is a call to end U.S. support for public financing of new coal-fired power plants overseas except for those employing advanced efficiency or carbon capture and sequestration technology. Notably, the plan provided no quantification of whether the United States would meet its commitment to reduce GHG emissions by 17% from 2005 levels by 2020 or whether or how the United States would work towards its share of a 2009 international pledge of $100 billion annually to assist developing countries to mitigate their GHG emissions and adapt to climate change.

PRESIDENTIAL ANNOUNCEMENT

On June 25, 2013, President Obama announced a national plan to reduce emissions of carbon dioxide (CO_2) and other greenhouse gases (GHG), as well as to encourage adaptation to expected climate change.[1] The President stated a willingness to work with Congress on a bipartisan, market-based scheme to reduce GHG emissions. However, the President had earlier stipulated that the set of actions he announced would not require Congressional approval. This announcement followed up on his vow in the 2013 State of the Union Address:

> I urge this Congress to pursue a bipartisan, market-based solution to climate change, like the one John McCain and Joe Lieberman worked on together a few years ago. But if Congress won't act soon to protect future generations, I will. I will direct my Cabinet to come up with executive actions we can take, now and in the future, to reduce pollution, prepare our communities for the consequences of climate change, and speed the transition to more sustainable sources of energy.[2]

The President had been under increasing pressure from environmental allies to exercise greater leadership on the climate change issue, after the Congress did not enact "Waxman-Markey" (H.R. 2454 in 2009) or other comprehensive bills to reduce GHG emissions. Also, some states and non-governmental organizations gave notice that they would file suit when the Environmental Protection Agency proposed but did not finalize greenhouse gas (GHG) emission standards for new power plants by April 2013.[3] The

President's plan is accompanied by a White House "infographic" covering extreme weather events, U.S. GHG emissions, and elements of the President's plan.[4]

Members of Congress continue to be divided in their views on whether climate change risks merit raising current costs to the economy in exchange for benefits that would mostly accrue to future generations, people in other countries, and stability of Earth systems. The prospect of continued congressional divisions in part has prompted the President's use of existing Executive Branch authorities.

PLEDGED ACTIONS AND TIMING

The President affirmed his commitment to his 2009 policy pledge to reduce U.S. GHG emissions by 17% below 2005 levels by 2020 if all other major economies agreed to limit their emissions as well.[5] In 2011, the United States' gross GHG emissions[6] were approximately 7% below their 2005 levels.[7] One study estimated in late 2012 that the United States is on a path to reduce its emissions to 16% below 2005 levels in 2020, under current policies including effects of the New Source Performance Standards for CO_2 emissions from power plants that the Environmental Protection Agency (EPA) proposed in 2012 (more below).[8] The researchers cite assumptions about promulgated and proposed standards, trends in the relative prices of coal and natural gas, and state and local GHG mitigation programs as main drivers of the projected reductions.

To fill the gap between the President's pledge and the current emissions trajectory, President Obama identified a series of actions to abate GHG emissions and to facilitate resilience to the effects of climate change. These are outlined below in three broad categories from the President's plan, which describes his three "pillars" as

1. Cut carbon pollution in America
2. Prepare the United States for the impacts of climate change
3. Lead international efforts to combat global climate change and prepare for its impacts.

This summary emphasizes the incremental additions in the plan announced June 25, 2013, beyond programs and other Executive Branch actions already underway. Few of the listed measures specify the timing of

Executive Branch actions, or the quantitative GHG reductions that should be achieved.

Cut Carbon Pollution

GHG Standards for Electricity Generation

The centerpiece of the President's announcement arguably is a Presidential Memorandum,[9] also issued June 25, that directs EPA to issue two types of rules to curtail carbon dioxide emissions from new and existing power plants before the end of his term.[10] Specifically, the Presidential Memorandum first instructs EPA to issue, as planned, a new proposal under the Clean Air Act (CAA), by September 20, 2013, for GHG emissions from newly constructed electric generating units (EGU), and to issue the final rule "in a timely fashion" after comments. Second, and most significantly, the Presidential Memorandum directs EPA also to issue standards, regulations, or guidelines for CO_2 emissions applicable to modified, reconstructed and existing power plants, building on States' efforts to reduce power plant emissions.

The Memorandum included neither specific levels of EGU emissions performance nor target-sectoral GHG reductions. The Memorandum requests the proposed rules for existing EGU by June 1, 2014, and final rules by June 1, 2015. Further, the Memorandum requests that the guidelines require States to submit to EPA their implementation plans, required under §111(d) of the CAA, and their implementing regulations by June 30, 2016. The Memorandum states that:

- states and non-governmental leaders should be directly engaged in the process;
- regulations and guidelines should be tailored to reduce costs;
- approaches should allow use of market-based instruments, performance standards and other flexibilities;
- standards should be consistent with maintaining reliable and affordable power; and
- EPA should work with states and other agencies to promote electricity from cleaner and more efficient technologies, including appliance efficiencies.

The President also directed EPA to work with state and local governments, industry, nongovernmental organizations, tribal officials and others in designing the programs, and to use market-based elements where possible. These instructions might be consistent with new standards and guidelines that take innovative forms, such as proposals put forward by policy analysts.[11] Some call for state-specific guidelines for existing power plants with flexibilities in state implementation plans to allow compliance in flexible ways, including use of generator and consumer efficiency, renewable mechanisms, rate averaging, and additional options. The President's announcement specifically indicated that the EPA standards should "provide flexibility to different states with different needs, and build on the leadership that many states, and cities, and companies have already shown."

Electricity Generation from Renewable Energy: The President set a goal to double electricity generation from wind, solar, and geothermal energy from current levels by 2020. Electricity generation produced from wind, solar, geothermal, increased 80% from 2009 to 2011, from 90 million megawatt-hours to 161 million megawatt-hours.[12] Including biomass fuels, the increase of non-hydro renewable electricity was 52% from 2009 to 2011, from 144 million megawatt-hours to 219 million megawatt-hours. The President proposes to accomplish the doubling goal by

- instructing the Department of Interior to issue permits for 10 gigawatts (GW) of renewable electric capacity on public lands by 2020;
- encouraging expansion of hydropower generation at existing dams, including the Red Rock Hydroelectric Plant on the Des Moines River in Iowa as a high priority permitting project.
- deploying 3 GW of renewable capacity on military installations by 2025;
- aiming to install 100 MW of renewable capacity for federally subsidized housing stock by 2020; and
- directing federal agencies, through a June 2013 Presidential Memorandum, to streamline siting, permitting, and review of electricity transmission projects across federal, state, and tribal government processes.

Consider Climate Impacts in the National Interests of Keystone XL: The State Department faces a pending decision of whether to grant a Presidential Permit for the proposed Keystone XL pipeline. Keystone XL would transport oil sands crude from Canada to a market hub in Nebraska for further delivery to Gulf Coast refineries.[13] Some environmental groups called this decision the "line in the sand" for U.S. climate change policy[14] and urged the President to deny the permit. On June 25, 2013, the President stated in his speech that "[our] national interest will be served only if this project does not significantly exacerbate the impacts of carbon pollution. The net effects of the pipeline's impact on our climate will be absolutely critical to determining whether this project can go forward."[15] In April 2013 comments on the draft Environmental Impact Statement (EIS) for the permit application, EPA recommended that the State Department include in the final EIS monetized estimates of the social cost of the GHG emissions from a barrel of oil sands crude compared to average U.S. crude. It is not clear whether the State Department will include in the final EIS the additional emissions data recommended by EPA. If those data are included, it is not yet known how it may inform the State Department's assessment of the "net effects" of the pipeline project on the climate.

Eliminate Tax Incentives Benefiting U.S. Fossil Fuels: The President proposed phasing out tax provisions that benefit fossil fuels in his FY2014 budget proposal.[16] (He links this proposal to seeking a global phase-out of similar incentives for fossil fuels.) The President proposed similar phase-outs in previous budget requests.

Technology Development: The President's FY2014 budget proposes approximately $7.9 billion, a 30% increase of funding across agencies, for research, development, and deployment of "clean energy" technologies.

- The Department of Energy (DOE) plans to issue a Notice in the *Federal Register* announcing a draft solicitation to use up to $8 billion in "Section 1703" loan guarantee authority[17] for advanced fossil fuel energy projects that can "cost-effectively meet financial and policy goals, including the avoidance, reduction, or sequestration of anthropogenic emissions of greenhouse gases."[18] The draft solicitation will be open for public comment, with an aim to issue the final solicitation in the Fall of 2013.
- Conduct a Federal Quadrennial Energy Review, led by the White House Domestic Policy Council and the Office of Science and Technology Policy, to assess energy infrastructure challenges, identify threats, risks, and opportunities for U.S. energy and climate security in order to translate policy goals into sequenced actions and proposed investments over four-year planning horizons.

Increasing Fuel Economy Standards for Heavy-Duty Vehicles for the post-2018 Model Years. No levels of performance or GHG reductions are specified in the plan. Model Year 2014- 2018 standards have been promulgated. Medium- and heavy-duty vehicles are the second largest component of transportation-sector CO_2 emissions, after light-duty vehicles (passenger cars and light trucks), while transportation emissions are 33% of U.S. CO_2 emissions and 28% of all U.S. GHG emissions.[19,20]

Developing and Deploying Next-Generation Energy Sources for Transportation: Continuing to support the Renewable Fuels Standard,[21] the Administration will invest in research and development on advanced biofuels, advanced batteries, and fuel cells in every mode of transportation. The Department of Transportation (DOT) will work with other agencies to explore how to integrate alternative fuel vessels into the U.S. flag fleet. DOT, the Department of Housing and Urban Development (HUD), and the EPA will work with states, cities, and towns to improve transportation options and lower transportation costs while protecting the environment.

New Goal for Energy Efficiency Standards: The President set a goal for efficiency standards for appliances and federal buildings, promulgated during his tenure, to reduce carbon emissions by at least 3 billion metric tons cumulatively by 2030 while also reducing household energy bills.

Department of Agriculture (USDA) Financing of Rural Energy Efficiency: USDA's Rural Utilities Service Energy Efficiency and Conservation Loan Program would provide up to $250 million for rural electric utilities to finance energy efficiency investments by businesses and households. The existing Rural Energy for America program would streamline grants and loan guarantees for efficiency and renewable energy investments by agricultural producers and rural small businesses.

Explore New Incentives for Residential Energy Efficiency: HUD's Multifamily Energy Innovation Fund would continue to provide $23 million to test new approaches to achieve cost-effective residential energy. The Federal Housing Administration with stakeholders will explore incentives in mortgage underwriting and appraisal that would factor energy efficiency into sales and refinancing of homes.

Expand DOE's Voluntary Better Buildings Challenge to Include Multifamily Housing: This program, which supports commercial and industrial building owners to improve energy efficiency by providing technical assistance and matching partners with allied suppliers, will expand to building owners and public housing agencies to cover multifamily housing efficiency as well.

Phase-Down Hydrofluorocarbon (HFC) Emissions: EPA will use its authority under the Significant New Alternatives Policy Program to identify and approve "climate –friendly" chemicals as alternatives to HFC and to chemicals that deplete the stratospheric ozone layer, and to prohibit certain uses of the most harmful chemicals. The Administration will purchase safer alternatives to HFC where feasible and transition to alternatives over time.

Develop an Interagency Methane Strategy: EPA, USDA, DOE, Department of Interior (DOI), Department of Labor (DOL) and DOT will assess best technologies and practices, and will identify existing authorities and incentives to reduce methane emissions.

Pursue Collaboration to Reduce Methane Emissions: The Administration will continue agencies' efforts with states, industrial sectors, and others, to reduce emissions, building on lessons learned from previous methane initiatives of former administrations.[22] The plan specifically mentions support for production of oil and gas while reducing venting and flaring of methane.

Protecting Carbon Removals by U.S. Resources and Landscapes: Related to measures to adapt to climate change, conservation and land management will seek to protect and restore forests, grasslands, wetlands, and other resources. In the context of a changing climate, these measures aim to help ensure that vegetation continues to remove carbon from the atmosphere, along with providing other services.

Continue to Carry Out Executive Order 13514: Agencies will continue to enter into performance-based contracts that promote energy savings[23] and take additional actions to promote energy efficiency. The plan indicates that agencies will synchronize building codes for federally owned and supported buildings, and take additional measures to meet the GHG emission reduction directives under E.O. 13514 (with adaptation plans discussed below).[24]

Prepare the United States for the Impacts of Climate Change

On October 5, 2009, President Obama signed Executive Order 13514, *Federal Leadership in Environmental, Energy, and Economic Performance*. The E.O. called for:

- federal agencies to participate actively to develop "approaches through which the policies and practices of the agencies can be made compatible with and reinforce" a national climate change adaptation strategy; and

- the Council on Environmental Quality to report to the President on progress on agency actions and recommendations for further measures.

Agencies were instructed by CEQ to include initial adaptation plans in their 2012 Strategic Sustainability Performance Plans (SSPPs), required of every agency under E.O. 13514. Agencies' SSPPs are available using sustainability.performance.gov. In many cases, the climate change adaptation plans are stand-alone reports appended to the SSPPs. The first adaptation plans were officially released on February 7, 2013, for a 60-day public comment period.

The June 2013 Climate Action Plan calls for a series of actions that may be—but are not clearly—incremental to what agencies were already undertaking pursuant to E.O. 13514. The plan includes:

- Directing agencies to identify and remove barriers to making climate-resilient investments; identify and remove counterproductive policies that increase vulnerabilities, and support more resilient investments through agency grants, technical assistance, and other mechanisms. Climate risk management is to be fully integrated into federal infrastructure and natural resource management planning,[25] the Clean Water and Drinking Water State Revolving Funds,[26] grants for brownfields cleanup, and HUD grants to assist in recovery following Superstorm Sandy.
- Establishing a State, Local, and Tribal Leaders Task Force on Climate Preparedness,[27] to provide recommendations to remove barriers to appropriate investments, modify grant and loan programs, and develop better information and tools to support communities that seek to become more resilient to a changing climate.
- Requiring that existing federal programs continue to provide targeted assistance to communities to prepare for the impacts of climate change, including through the Federal Highway Administration, the Bureau of Indian Affairs, and annual federal "Environmental Justice Progress Reports."
- Boosting resilience of buildings and infrastructure by developing a framework and guidelines for safe buildings and infrastructure through a panel to be convened by the National Institute of Standards and Technology (NIST). The President's FY2014 budget proposal

included $200 million for Climate Ready Infrastructure through the Transportation Leadership Awards program of DOT.

- Enhancing resilience in rebuilding following Hurricane Sandy through federal relief programs. Programs that provide financial assistance include the Federal Transit Administration ($1.3 billion to locally prioritized projects to make transit systems more resilient to future disasters); DOI ($100 million in competitive grants to promote resilient natural systems and $250 million in support projects for coastal restoration and resilience); and the U.S. Army Corps of Engineers ($20 million to study reducing the vulnerability of Sandy-affected coastal communities to future large-scale flood and storm events);.[28]
- Identifying and taking actions in specific sectors, including energy production, healthcare, insurance, land and water resource conservation and management;agriculture, forestry, and others. Among these efforts would be launch of a new National Drought Resilience Partnership, and expansion of forest- and rangeland-restoration to reduce those areas' vulnerability to catastrophic fire.
- Focusing on "usable knowledge" in continuing global change research and completing the third National Climate Assessment (NCA). The expected release date for the NCA is now Spring of 2014.
- Launching a Climate Data Initiative consistent with the President's May 9, 2013, Executive Order, *Making Open and Machine Readable the New Default for Government Information*.[29] It is intended to make government data more freely available.
- Providing a Toolkit for Climate Resilience that centralizes access to various federal agencies' related tools, services, and best practices.

International Leadership

When the United States ratified the 1992 United Nations Framework Convention on Climate Change (UNFCCC), it agreed to the objective of

> ...stabilization of greenhouse gas concentrations in the atmosphere at a level that would prevent dangerous anthropogenic interference with the climate system. Such a level should be achieved with a time-frame sufficient to allow ecosystems to adapt naturally to climate change, to

ensure that food production is not threatened and to enable economic development to proceed in a sustainable manner.[30]

Although the United States signed the 1997 Kyoto Protocol to the UNFCCC, the agreement was never submitted to the Senate for consent to its ratification.[31] Congressional opposition was strong to a treaty, in part because it did not include all major emitting countries, such as China, and that it would impose costs on the U.S. economy.[32] The United States has no quantitative, legally binding obligations to reduce its GHG emissions under the UNFCCC, although it is currently negotiating towards an agreement to abate GHG emissions globally, due in 2015 and to take effect by 2020.

After President Obama took office, in conjunction with the 2009 Copenhagen Accord, he pledged a policy to reduce U.S. GHG emissions by 17% compared with 2005 levels. Without rules to reduce power plant emissions, the United States would not be on track to meet this pledge, a fact that increases pressure and arguably undermines U.S. leadership (though not necessarily leverage) in the international negotiations.

New in the President's announcement is a call to end U.S. support for public financing of new coal-fired power plants overseas. This would exempt the most efficient coal technology available in the world's poorest countries and power plants that employ emerging carbon capture and sequestration technology.

The President's Climate Action Plan includes several efforts in motion prior to the June 2013 announcements. Among the bilateral and multilateral programs are:

- A "major initiative" in 2013 in conjunction with the Major Economies Forum on Energy and Climate[33] to promote energy efficiency gains in buildings.
- Bilateral cooperation with key major emerging economies. Existing initiatives include the U.S.-China Clean Energy Research Center, the U.S.-India Partnership to Advance Clean Energy, and the Strategic Energy Dialogue with Brazil. The President cites as an example, an agreement between the United States and China in June 2013 to phase down production and consumption of hydrofluorocarbons (HFC) globally under the Montreal Protocol on Substances that Deplete the Ozone Layer.[34,35]
- A call to phase out subsidies for fossil fuels globally. G-20 leaders committed to this goal in 2009. President Obama proposes in his

FY2014 budget, as in previous budgets, to eliminate U.S. tax provisions that benefit fossil fuel supply.

- Negotiating a global free trade agreement on environmental goods and services under the World Trade Organization (WTO). The President said such a global agreement would build on the 2011 agreement under the Asia-Pacific Economic Cooperation (APEC) economies to reduce tariffs to 5% or less by 2015 on a negotiated list of 54 environmental goods. The plan calls for negotiation at the WTO of a plurilateral agreement to eliminate tariffs on environmental goods with countries that account for 90% of global trade in environmental goods. It also calls for inclusion of environmental services in the ongoing plurilateral Trade in International Services agreement. These talks seek to supplant the thus far unsuccessful multilateral negotiations on environmental goods and services at the WTO Doha Round.
- Supporting reduction of emissions from deforestation and forest degradation (REDD) globally. The Agency for International Development (AID) works bilaterally and multilaterally through the Forest Investment Program, the Forest Carbon Partnership Facility, the Millennium Challenge Corporation, and the Tropical Forest Alliance 2020.
- Supporting international cooperation to reduce so-called "super-pollutants", gases and aerosol pollutants that are highly potent but short-lived once emitted to, or formed from emissions in, the atmosphere. The Climate and Clean Air Coalition to Reduce Short-Lived Climate Pollution was formed among a group of like-minded countries in February 2012 to promote reduction of methane, black carbon, HFC, and other pollutants with potent influence on the climate.

In cooperation under the UNFCCC, the Obama Administration pledged to help mobilize $30 billion of financing during 2010-2012 to assist developing countries mitigate GHG emissions and adapt to climate change. The $30 billion was intended as a publicly funded "fast start" toward stimulating financing – both public and private – of $100 billion annually by 2020. The President stated the United States met its pledge for the 2010-2012 "fast-start" financing with approximately $7.5 billion in that three-year period.[36] The June 2013 Plan speaks to the pledge for 2020 financing in general terms only.[37]

POSSIBLE ISSUES FOR CONGRESS

The President's plan mostly envisions actions that do not require further authority from Congress to accomplish, although some would rely on appropriations of funds to specific programs. An initial review of the Climate Action Plan raises a few initial questions some in Congress may wish to explore.

1. The cumulative impacts of existing measures and the President's announced plan have not been assessed, in terms of accomplishments in reducing GHG emissions or in terms of costs and savings that may result. In addition to net costs or benefits, distributional impacts across regions, sectors, and segments of the population may be of interest.
2. Some in Congress and, more generally, in the American public oppose regulation of GHG emissions under the Clean Air Act. Some legislators have proposed resolutions or bills to curtail executive authority in advance of EPA finalization of rules. Once a rule is finalized, Congress may use the Congressional Review Act (CRA, 5 U.S.C. §§ 801 *et seq.*)[38] to overturn the rule by enacting a joint resolution of disapproval. The CRA allows for the use of expedited procedures when considering such a joint resolution, and, if the joint resolution were to be enacted, the rule would no longer have force or effect. However, the resolution of disapproval would require the signature of the President, who would be unlikely to sign a bill overturning his own agency's rule. A two-thirds vote in both House and Senate could overcome a Presidential veto.

   In May 2013, the Administration released an updated range for the "social costs of carbon" (SCC)—dollar values representing the benefits of avoiding damages of climate change, expressed as dollars per ton of CO_2 released in particular years.[39] These SCC estimates have been, and will be used, to compare the benefits with the costs of prospective regulations,[40] though cost-benefit analysis has not typically been a determinative factor in regulatory choices. While the interagency group that developed the SCC acknowledged the difficulties of providing such values, the 9th Circuit Court ruled in 2007 that estimating the benefits of rules without including such values for climate damages is "arbitrary and capricious."[41] Some observers have expressed concern that the new SCC values could be used to justify tighter standards, while others have criticized the methods for understating the SCC. Congress may wish to consider the process by which the SCC have been developed, the role of public and Congressional input, and how the SCC may affect future actions.

3. Some in Congress may seek to enact a more cohesive alternative to EPA regulation of GHG emissions and the assortment of administration measures in the Obama plan. Some policy-makers have proposed that fees on GHG emissions (e.g., "carbon taxes") would be a preferable alternative, while others oppose this option. The Keystone XL pipeline, to transport Canadian oil sands crude across the U.S. border, requires a Presidential Permit from the President. While many support approval of the pipeline, some environmental proponents have called the project the "line in the sand" on climate change and urge the President to deny the permit. Pursuant to the President's statement that "[our] national interest will be served only if this project does not significantly exacerbate the impacts of carbon pollution," Congress may wish to examine systematically the trade-offs to national interest of the project.

End Notes

[1] Executive Office of the President (EOP). The President's Climate Action Plan. June 2013. Available at http://www.whitehouse.gov/sites/default/files/image/president27 sclimate actionplan.pdf.

[2] White House, Remarks by the President in the State of the Union Address. Washington DC. February 12, 2013.

[3] The House passed the American Clean Energy and Security Act (the "Waxman-Markey" bill), the 111th Congress' H.R. 2454, but no corresponding bill cleared the Senate. For further information, see CRS Report R40643, Greenhouse Gas Legislation: Summary and Analysis of H.R. 2454 as Passed by the House of Representatives, coordinated by Mark Holt and Gene Whitney.

[4] White House. President Obama's Plan to Fight Climate Change. http://www.whitehouse.gov/share/climate-actionplan.

[5] President Obama separately set a goal to double U.S. energy productivity (i.e., energy consumption per unit of economic activity, such as Gross Domestic Product) by 2030 compared with 2010 levels.

[6] Of the six gases covered by the Kyoto Protocol, excluding hydrofluorocarbons, nitrogen hexafluoride and other newer GHG, and excluding removals of CO2 from the atmosphere by "sinks", such as uptake by expanding forests.

[7] CRS calculations from data in Environmental Protection Agency. Inventory of U.S. Greenhouse Gas Emission and Sinks: 1990-2011. Washington DC, April 2013. http://www.epa.gov/climatechange/ghgemissions/ usinventoryreport.html.

[8] Burtraw, Dallas, and Matt Woerman. US Status on Climate Change Mitigation. Washington DC: Resources for the Future, October 2012. http://www.rff.org/RFF/Documents/RFF-DP-12-48.pdf.

[9] White House. "Memorandum for the Administrator of the Environmental Protection Agency: Power Sector Carbon Pollution Standards." June 25, 2013.

[10] For more information on these rules, see CRS Report R43127, EPA Standards for Greenhouse Gas Emissions from Power Plants: Many Questions, Some Answers, by James E. McCarthy.

[11] See, for example, Tarr, Jeremy et al., Regulating CO2 under Section 111(d): Options, Limits, and Impacts. Nicholas Institute for Environmental Policy Solutions, Duke University. January 2013; and Lashof, Daniel A., et al. Closing the Power Plant Carbon Pollution Loophole: Smart Ways the Clean Air Act Can Clean Up America's Biggest Climate Polluters. Natural Resources Defense Council. 2012.

[12] Energy Information Administration. Net Generation by Other Renewable Sources: Total (All Sectors), 2003-April 2013. Extracted June 25, 2013. Data available at http://www.eia.gov/electricity/monthly/epm_table_grapher.cfm?t= epmt_1_01_a.

[13] See CRS Report R41668, Keystone XL Pipeline Project: Key Issues, by Paul W. Parfomak et al.

[14] For analysis of the emissions implications of the Keystone pipelines, see CRS Report R42611, Oil Sands and the Keystone XL Pipeline: Background and Selected Environmental Issues, coordinated by Jonathan L. Ramseur.

[15] White House. Remarks by the President on Climate Change. (transcript) Georgetown University, Washington DC. June 25, 2013. http://www.whitehouse.gov/the-press-office/2013/06/25/remarks-president-climate-change.

[16] For more on this topic in the FY2013 budget proposal, see CRS Report R42374, Oil and Natural Gas Industry Tax Issues in the FY2013 Budget Proposal, by Robert Pirog.

[17] Energy Policy Act (EPAct) of 2005, Title XVII, §1703.

[18] EOP, op. cit., p. 7.

[19] EPA, op.cit.

[20] See CRS Report R40506, Cars, Trucks, and Climate: EPA Regulation of Greenhouse Gases from Mobile Sources, by James E. McCarthy and Brent D. Yacobucci.

[21] See CRS Report R40155, Renewable Fuel Standard (RFS): Overview and Issues, by Randy Schnepf and Brent D. Yacobucci.

[22] See CRS Report R40813, Methane Capture: Options for Greenhouse Gas Emission Reduction, by Kelsi Bracmort et al.

[23] Directed by Presidential Memorandum, "Implementation of Energy Savings Projects and Performance-Based Contracting for Energy Savings." December 2, 2011.

[24] http://www.whitehouse.gov/administration/eop/ceq/sustainability.

[25] See GAO. Climate Change: Various Adaptation Efforts Are Under Way at Key Natural Resource Management Agencies, May 31, 2013. http://www.gao.gov/products/GAO-13-253.

[26] For more about these funds, see CRS Report RL31116, Water Infrastructure Needs and Investment: Review and Analysis of Key Issues, by Claudia Copeland and Mary Tiemann.

[27] See GAO. Climate Change: Future Federal Adaptation Efforts Could Better Support Local Infrastructure Decision Makers, April 12, 2013. http://www.gao.gov/products/GAO-13-242.

[28] EOP, op.cit., p. 14.

[29] White House. Making Open and Machine Readable the New Default for Government Information. May 9, 2013.

[30] UNFCCC Article 2, Objective. 1992.

[31] For more information on the international negotiations on climate change, see CRS Report R40001, A U.S.-Centric Chronology of the International Climate Change Negotiations, by Jane A. Leggett.

[32] The Senate had stated this clearly, in July 2007 prior to the Kyoto negotiations, in S.Res. 98, passed on a vote of 95- 0.

[33] http://www.state.gov/e/oes/climate/mem/index.htm.

[34] For more information on the Montreal Protocol, see United Nations Environment Programme, Ozone Secretariat. The Montreal Protocol on Substances that Deplete the Ozone Layer.

[35] White House. United States and China Agree to Work Together on Phase Down of HFCs. Press release. June 8, 2013.

[36] The State Department's calculation is available at http://www.state.gov/e/oes/climate/faststart/index.htm.

[37] EOP. op.cit., p. 20.

[38] CRS Report RL32240, The Federal Rulemaking Process: An Overview, coordinated by Maeve P. Carey.

[39] InteragencyWorking Group on Social Cost of Carbon. Technical Support Document: Technical Update of the Social Cost of Carbon for Regulatory Impact Analysis Under Executive Order 12866. Washington DC: U.S. Government, May 2013. http://www.whitehouse.gov/sites/default/files/omb/inforeg/social_cost_of_carbon_for_ria_2013_update.pdf.

[40] CRS Report R41974, Cost-Benefit and Other Analysis Requirements in the Rulemaking Process, by Maeve P. Carey.

[41] Center for Biological Diversity v. National Highway Traffic Safety Admin., 508 F.3d 508, 519 (9th Cir. 2007).

In: President Obama's Climate Action Plan ISBN: 978-1-62808-863-2
Editor: Shane N. Blake © 2013 Nova Science Publishers, Inc.

Chapter 2

THE PRESIDENT'S CLIMATE ACTION PLAN*

Executive Office of the President

PRESIDENT OBAMA'S CLIMATE ACTION PLAN

"We, the people, still believe that our obligations as Americans are not just to ourselves, but to all posterity. We will respond to the threat of climate change, knowing that the failure to do so would betray our children and future generations. Some may still deny the overwhelming judgment of science, but none can avoid the devastating impact of raging fires and crippling drought and more powerful storms.

The path towards sustainable energy sources will be long and sometimes difficult. But America cannot resist this transition, we must lead it. We cannot cede to other nations the technology that will power new jobs and new industries, we must claim its promise. That's how we will maintain our economic vitality and our national treasure -- our forests and waterways, our croplands and snow-capped peaks. That is how we will preserve our planet, commanded to our care by God. That's what will lend meaning to the creed our fathers once declared."

-- President Obama, Second Inaugural Address, January 2013

---

* This is an edited, reformatted and augmented version of an Executive Office of the President publication, dated June 2013.

THE CASE FOR ACTION

While no single step can reverse the effects of climate change, we have a moral obligation to future generations to leave them a planet that is not polluted and damaged. Through steady, responsible action to cut carbon pollution, we can protect our children's health and begin to slow the effects of climate change so that we leave behind a cleaner, more stable environment.

In 2009, President Obama made a pledge that by 2020, America would reduce its greenhouse gas emissions in the range of 17 percent below 2005 levels if all other major economies agreed to limit their emissions as well. Today, the President remains firmly committed to that goal and to building on the progress of his first term to help put us and the world on a sustainable long-term trajectory. Thanks in part to the Administration's success in doubling America's use of wind, solar, and geothermal energy and in establishing the toughest fuel economy standards in our history, we are creating new jobs, building new industries, and reducing dangerous carbon pollution which contributes to climate change. In fact, last year, carbon emissions from the energy sector fell to the lowest level in two decades. At the same time, while there is more work to do, we are more energy secure than at any time in recent history. In 2012, America's net oil imports fell to the lowest level in 20 years and we have become the world's leading producer of natural gas – the cleanest-burning fossil fuel.

While this progress is encouraging, climate change is no longer a distant threat – we are already feeling its impacts across the country and the world. Last year was the warmest year ever in the contiguous United States and about one-third of all Americans experienced 10 days or more of 100-degree heat. The 12 hottest years on record have all come in the last 15 years. Asthma rates have doubled in the past 30 years and our children will suffer more asthma attacks as air pollution gets worse. And increasing floods, heat waves, and droughts have put farmers out of business, which is already raising food prices dramatically.

These changes come with far-reaching consequences and real economic costs. Last year alone, there were 11 different weather and climate disaster events with estimated losses exceeding $1 billion each across the United States. Taken together, these 11 events resulted in over $110 billion in estimated damages, which would make it the second-costliest year on record.

In short, America stands at a critical juncture. Today, President Obama is putting forward a broad-based plan to cut the carbon pollution that causes climate change and affects public health. Cutting carbon pollution will help spark business innovation to modernize our power plants, resulting in cleaner forms of American-made energy that will create good jobs and cut our dependence on foreign oil. Combined with the Administration's other actions to increase the efficiency of our cars and household appliances, the President's plan will reduce the amount of energy consumed by American families, cutting down on their gas and utility bills. The plan, which consists of a wide variety of executive actions, has three key pillars:

1. *Cut Carbon Pollution in America:* In 2012, U.S. carbon emissions fell to the lowest level in two decades even as the economy continued to grow. To build on this progress, the Obama Administration is putting in place tough new rules to cut carbon pollution – just like we have for other toxins like mercury and arsenic – so we protect the health of our children and move our economy toward American-made clean energy sources that will create good jobs and lower home energy bills.
2. *Prepare the United States for the Impacts of Climate Change:* Even as we take new steps to reduce carbon pollution, we must also prepare for the impacts of a changing climate that are already being felt across the country. Moving forward, the Obama Administration will help state and local governments strengthen our roads, bridges, and shorelines so we can better protect people's homes, businesses and way of life from severe weather.
3. *Lead International Efforts to Combat Global Climate Change and Prepare for its Impacts:* Just as no country is immune from the impacts of climate change, no country can meet this challenge alone. That is why it is imperative for the United States to couple action at home with leadership internationally. America must help forge a truly global solution to this global challenge by galvanizing international action to significantly reduce emissions (particularly among the major emitting countries), prepare for climate impacts, and drive progress through the international negotiations.

Climate change represents one of our greatest challenges of our time, but it is a challenge uniquely suited to America's strengths. Our scientists will design new fuels, and our farmers will grow them. Our engineers to devise new sources of energy, our workers will build them, and our businesses will sell them. All of us will need to do our part. If we embrace this challenge, we will not just create new jobs and new industries and keep America on the cutting edge; we will save lives, protect and preserve our treasured natural resources, cities, and coastlines for future generations.

What follows is a blueprint for steady, responsible national and international action to slow the effects of climate change so we leave a cleaner, more stable environment for future generations. It highlights progress already set in motion by the Obama Administration to advance these goals and sets forth new steps to achieve them.

Cut Carbon Pollution in America

In 2009, President Obama made a commitment to reduce U.S. greenhouse gas emissions in the range of 17 percent below 2005 levels by 2020. The President remains firmly committed to achieving that goal. While there is more work to do, the Obama Administration has already made significant progress by doubling generation of electricity from wind, solar, and geothermal, and by establishing historic new fuel economy standards. Building on these achievements, this document outlines additional steps the Administration will take – in partnership with states, local communities, and the private sector – to continue on a path to meeting the President's 2020 goal.

I. Deploying Clean Energy

Cutting Carbon Pollution from Power Plants: Power plants are the largest concentrated source of emissions in the United States, together accounting for roughly one-third of all domestic greenhouse gas emissions. We have already set limits for arsenic, mercury, and lead, but there is no federal rule to prevent power plants from releasing as much carbon pollution as they want. Many states, local governments, and companies have taken steps to move to cleaner electricity sources. More than 35 states have renewable energy targets in place, and more than 25 have set energy efficiency targets.

Despite this progress at the state level, there are no federal standards in place to reduce carbon pollution from power plants. In April 2012, as part of a continued effort to modernize our electric power sector, the Obama Administration proposed a carbon pollution standard for new power plants. The Environmental Protection Agency's proposal reflects and reinforces the ongoing trend towards cleaner technologies, with natural gas increasing its share of electricity generation in recent years, principally through market forces and renewables deployment growing rapidly to account for roughly half of new generation capacity installed in 2012.

With abundant clean energy solutions available, and building on the leadership of states and local governments, we can make continued progress in reducing power plant pollution to improve public health and the environment while supplying the reliable, affordable power needed for economic growth. By doing so, we will continue to drive American leadership in clean energy technologies, such as efficient natural gas, nuclear, renewables, and clean coal technology.

To accomplish these goals, President Obama is issuing a Presidential Memorandum directing the Environmental Protection Agency to work expeditiously to complete carbon pollution standards for both new and existing power plants. This work will build on the successful first-term effort to develop greenhouse gas and fuel economy standards for cars and trucks. In developing the standards, the President has asked the Environmental Protection Agency to build on state leadership, provide flexibility, and take advantage of a wide range of energy sources and technologies including many actions in this plan.

Promoting American Leadership in Renewable Energy: During the President's first term, the United States more than doubled generation of electricity from wind, solar, and geothermal sources. To ensure America's continued leadership position in clean energy, President Obama has set a goal to double renewable electricity generation once again by 2020. In order to meet this ambitious target, the Administration is announcing a number of new efforts in the following key areas:

- *Accelerating Clean Energy Permitting:* In 2012 the President set a goal to issue permits for 10 gigawatts of renewables on public lands by the end of the year. The Department of the Interior achieved this goal ahead of schedule and the President has directed it to permit an additional 10 gigawatts by 2020. Since 2009, the Department of Interior has approved 25 utility-scale solar facilities, nine wind farms, and 11 geothermal plants, which will provide enough electricity to power 4.4 million homes and support an estimated 17,000 jobs. The Administration is also taking steps to encourage the development of hydroelectric power at existing dams. To develop and demonstrate improved permitting procedures for such projects, the Administration will designate the Red Rock Hydroelectric Plant on the Des Moines River in Iowa to participate in its Infrastructure Permitting Dashboard for high-priority projects. Also, the Department of Defense – the single largest consumer of energy in the United States – is committed to deploying 3 gigawatts of renewable energy on military installations, including solar, wind, biomass, and geothermal, by 2025. In addition, federal agencies are setting a new goal of reaching 100 megawatts of installed renewable capacity across the federally subsidized housing stock by 2020. This effort will include conducting a survey of current projects in order to track progress and facilitate the sharing of best practices.

- *Expanding and Modernizing the Electric Grid:* Upgrading the country's electric grid is critical to our efforts to make electricity more reliable, save consumers money on their energy bills, and promote clean energy sources. To advance these important goals, President Obama signed a Presidential Memorandum this month that directs federal agencies to streamline the siting, permitting and review process for transmission projects across federal, state, and tribal governments.

Unlocking Long-Term Investment in Clean Energy Innovation: The Fiscal Year 2014 Budget continues the President's commitment to keeping the United States at the forefront of clean energy research, development, and deployment by increasing funding for clean energy technology across all agencies by 30 percent, to approximately $7.9 billion. This includes investment in a range of energy technologies, from advanced biofuels and emerging nuclear technologies – including small modular reactors – to clean coal. To continue America's leadership in clean energy innovation, the Administration will also take the following steps:

- *Spurring Investment in Advanced Fossil Energy Projects*: In the coming weeks, the Department of Energy will issue a Federal Register Notice announcing a draft of a solicitation that would make up to $8 billion in (self-pay) loan guarantee authority available for a wide array of advanced fossil energy projects under its Section 1703 loan guarantee program. This solicitation is designed to support investments in innovative technologies that can cost-effectively meet financial and policy goals, including the avoidance, reduction, or sequestration of anthropogenic emissions of greenhouse gases. The proposed solicitation will cover a broad range of advanced fossil energy projects. Reflecting the Department's commitment to continuous improvement in program management, it will take comment on the draft solicitation, with a plan to issue a final solicitation by the fall of 2013.

- *Instituting a Federal Quadrennial Energy Review*: Innovation and new sources of domestic energy supply are transforming the nation's energy marketplace, creating economic opportunities at the same time they raise environmental challenges. To ensure that federal energy policy meets our economic, environmental, and security goals in this changing landscape, the Administration will conduct a Quadrennial Energy Review which will be led by the White House Domestic Policy Council and Office of Science and Technology Policy, supported by a Secretariat established at the Department of Energy, and involving the robust engagement of federal agencies and outside stakeholders. This first-ever review will focus on infrastructure challenges, and will identify the threats, risks, and opportunities for U.S. energy and climate security, enabling the federal government to translate policy goals into a set of analytically based, clearly articulated, sequenced and integrated actions, and proposed investments over a four-year planning horizon.

II. Building a 21st-Century Transportation Sector

Increasing Fuel Economy Standards: Heavy-duty vehicles are currently the second largest source of greenhouse gas emissions within the transportation sector. In 2011, the Obama Administration finalized the first-ever fuel economy standards for Model Year 2014-2018 for heavy-duty trucks, buses, and vans. These standards will reduce greenhouse gas emissions by approximately 270 million metric tons and save 530 million barrels of oil. During the President's second term, the Administration will once again partner with industry leaders and other key stakeholders to develop post-2018 fuel economy standards for heavy-duty vehicles to further reduce fuel consumption through the application of advanced cost-effective technologies and continue efforts to improve the efficiency of moving goods across the United States.

The Obama Administration has already established the toughest fuel economy standards for passenger vehicles in U.S. history. These standards require an average performance equivalent of 54.5 miles per gallon by 2025, which will save the average driver more than $8,000 in fuel costs over the lifetime of the vehicle and eliminate six billion metric tons of carbon pollution – more than the United States emits in an entire year.

Developing and Deploying Advanced Transportation Technologies: Biofuels have an important role to play in increasing our energy security, fostering rural economic development, and reducing greenhouse gas emissions from the transportation sector. That is why the Administration supports the Renewable Fuels Standard, and is investing in research and development to help bring next-generation biofuels on line. For example, the United States Navy and Departments of Energy and Agriculture are working with the private sector to accelerate the development of cost-competitive advanced biofuels for use by the military and commercial sectors. More broadly, the Administration will continue to leverage partnerships between the private and public sectors to deploy cleaner fuels, including advanced batteries and fuel cell technologies, in every transportation mode. The Department of Energy's eGallon informs drivers about electric car operating costs in their state – the national average is only $1.14 per gallon of gasoline equivalent, showing the promise for consumer pocketbooks of electric-powered vehicles. In addition, in the coming months, the Department of Transportation will work with other agencies to further explore strategies for integrating alternative fuel vessels into the U.S. flag fleet. Further, the Administration will continue to work with states, cities and towns through the Department of Transportation, the Department of Housing and Urban Development, and the Environmental Protection Agency to improve transportation options, and lower transportation costs while protecting the environment in communities nationwide.

III. Cutting Energy Waste in Homes, Businesses, and Factories

Reducing Energy Bills for American Families and Businesses: Energy efficiency is one of the clearest and most cost-effective opportunities to save families money, make our businesses more competitive, and reduce greenhouse gas emissions. In the President's first term, the Department of Energy and the Department of Housing and Urban Development completed efficiency upgrades in more than one million homes, saving many families more than $400 on their heating and cooling bills in the first year alone. The Administration will take a range of new steps geared towards achieving President Obama's goal of doubling energy productivity by 2030 relative to 2010 levels:

- *Establishing a New Goal for Energy Efficiency Standards*: In President Obama's first term, the Department of Energy established new minimum efficiency standards for dishwashers, refrigerators, and many other products. Through 2030, these standards will cut consumers' electricity bills by hundreds of billions of dollars and save enough electricity to power more than 85 million homes for two years. To build on this success, the Administration is setting a new goal: Efficiency standards for appliances and federal buildings set in the first and second terms combined will reduce carbon pollution by at least 3 billion metric tons cumulatively by 2030 – equivalent to nearly one-half of the carbon pollution from the entire U.S. energy sector for one year – while continuing to cut families' energy bills.
- *Reducing Barriers to Investment in Energy Efficiency*: Energy efficiency upgrades bring significant cost savings, but upfront costs act as a barrier to more widespread investment. In response, the Administration is committing to a number of new executive actions. As soon as this fall, the Department of Agriculture's Rural Utilities Service will finalize a proposed update to its Energy Efficiency and Conservation Loan Program to provide up to $250 million for rural utilities to finance efficiency investments by businesses and homeowners across rural America. The Department is also streamlining its Rural Energy for America program to provide grants and loan guarantees directly to agricultural producers and rural small businesses for energy efficiency and renewable energy systems.

  In addition, the Department of Housing and Urban Development's efforts include a $23 million Multifamily Energy Innovation Fund designed to enable affordable housing providers, technology firms, academic institutions, and philanthropic organizations to test new approaches to deliver cost-effective residential energy. In order to advance ongoing efforts and bring stakeholders together, the Federal Housing Administration will convene representatives of the lending community and other key stakeholders for a mortgage roundtable in July to identify options for factoring energy efficiency into the mortgage underwriting and appraisal process upon sale or refinancing of new or existing homes.
- *Expanding the President's Better Buildings Challenge*: The Better Buildings Challenge, focused on helping American commercial and industrial buildings become at least 20 percent more energy efficient by 2020, is already showing results. More than 120 diverse

organizations, representing over 2 billion square feet are on track to meet the 2020 goal: cutting energy use by an average 2.5 percent annually, equivalent to about $58 million in energy savings per year. To continue this success, the Administration will expand the program to multifamily housing – partnering both with private and affordable building owners and public housing agencies to cut energy waste. In addition, the Administration is launching the Better Buildings Accelerators, a new track that will support and encourage adoption of State and local policies to cut energy waste, building on the momentum of ongoing efforts at that level.

IV. Reducing Other Greenhouse Gas Emissions

Curbing Emissions of Hydrofluorocarbons: Hydrofluorocarbons (HFCs), which are primarily used for refrigeration and air conditioning, are potent greenhouse gases. In the United States, emissions of HFCs are expected to nearly triple by 2030, and double from current levels of 1.5 percent of greenhouse gas emissions to 3 percent by 2020.

To reduce emissions of HFCs, the United States can and will lead both through international diplomacy as well as domestic actions. In fact, the Administration has already acted by including a flexible and powerful incentive in the fuel economy and carbon pollution standards for cars and trucks to encourage automakers to reduce HFC leakage and transition away from the most potent HFCs in vehicle air conditioning systems. Moving forward, the Environmental Protection Agency will use its authority through the Significant New Alternatives Policy Program to encourage private sector investment in low-emissions technology by identifying and approving climate-friendly chemicals while prohibiting certain uses of the most harmful chemical alternatives. In addition, the President has directed his Administration to purchase cleaner alternatives to HFCs whenever feasible and transition over time to equipment that uses safer and more sustainable alternatives.

Reducing Methane Emissions: Curbing emissions of methane is critical to our overall effort to address global climate change. Methane currently accounts for roughly 9 percent of domestic greenhouse gas emissions and has a global warming potential that is more than 20 times greater than carbon dioxide. Notably, since 1990, methane emissions in the United States have decreased by 8 percent. This has occurred in part through partnerships with industry, both at home and abroad, in which we have demonstrated that we have the technology to deliver emissions reductions that benefit both our economy and the environment. To achieve additional progress, the Administration will:

- *Developing an Interagency Methane Strategy:* The Environmental Protection Agency and the Departments of Agriculture, Energy, Interior, Labor, and Transportation will develop a comprehensive, interagency methane strategy. The group will focus on assessing current emissions data, addressing data gaps, identifying technologies and best practices for reducing emissions, and identifying existing authorities and incentive-based opportunities to reduce methane emissions.
- *Pursuing a Collaborative Approach to Reducing Emissions:* Across the economy, there are multiple sectors in which methane emissions can be reduced, from coal mines and landfills to agriculture and oil and gas development. For example, in the agricultural sector, over the last three years, the Environmental Protection Agency and the Department of Agriculture have worked with the dairy industry to increase the adoption of methane digesters through loans, incentives, and other assistance. In addition, when it comes to the oil and gas sector, investments to build and upgrade gas pipelines will not only put more Americans to work, but also reduce emissions and enhance economic productivity. For example, as part of the Administration's effort to improve federal permitting for infrastructure projects, the interagency Bakken Federal Executive Group is working with industry, as well as state and tribal agencies, to advance the production of oil and gas in the Bakken while helping to reduce venting and flaring. Moving forward, as part of the effort to develop an interagency methane strategy, the Obama Administration will work collaboratively with state governments, as well as the private sector, to reduce emissions across multiple sectors, improve air quality, and achieve public health and economic benefits.

Preserving the Role of Forests in Mitigating Climate Change: America's forests play a critical role in addressing carbon pollution, removing nearly 12 percent of total U.S. greenhouse gas emissions each year. In the face of a changing climate and increased risk of wildfire, drought, and pests, the capacity of our forests to absorb carbon is diminishing. Pressures to develop forest lands for urban or agricultural uses also contribute to the decline of forest carbon sequestration. Conservation and sustainable management can help to ensure our forests continue to remove carbon from the atmosphere while also improving soil and water quality, reducing wildfire risk, and otherwise managing forests to be more resilient in the fact of climate change. The Administration is working to identify new approaches to protect and restore our forests, as well as other critical landscapes including grasslands and wetlands, in the face of a changing climate.

V. Leading at the Federal Level

Leading in Clean Energy: President Obama believes that the federal government must be a leader in clean energy and energy efficiency. Under the Obama Administration, federal agencies have reduced greenhouse gas emissions by more than 15 percent – the equivalent of permanently taking 1.5 million cars off the road. To build on this record, the Administration is establishing a new goal: The federal government will consume 20 percent of its electricity from renewable sources by 2020 – more than double the current goal of 7.5 percent. In addition, the federal government will continue to pursue greater energy efficiency that reduces greenhouse gas emissions and saves taxpayer dollars.

Federal Government Leadership in Energy Efficiency: On December 2, 2011, President Obama signed a memorandum entitled "Implementation of Energy Savings Projects and Performance-Based Contracting for Energy Savings," challenging federal agencies, in support of the Better Buildings Challenge, to enter into $2 billion worth of performance-based contracts within two years. Performance contracts drive economic development, utilize private sector innovation, and increase efficiency at minimum costs to the taxpayer, while also providing long- term savings in energy costs. Federal agencies have committed to a pipeline of nearly $2.3 billion from over 300 reported projects. In coming months, the Administration will take a number of actions to strengthen efforts to promote energy efficiency, including through performance contracting. For example, in order to increase access to capital markets for investments in energy efficiency, the Administration will initiate a partnership with the private sector to work towards a standardized contract to finance federal investments in energy efficiency. Going forward, agencies will also work together to synchronize building codes – leveraging those policies to improve the efficiency of federally owned and supported building stock. Finally, the Administration will leverage the "Green Button" standard – which aggregates energy data in a secure, easy to use format – within federal facilities to increase their ability to manage energy consumption, reduce greenhouse gas emissions, and meet sustainability goals.

PREPARE THE UNITED STATES FOR THE IMPACTS OF CLIMATE CHANGE

As we act to curb the greenhouse gas pollution that is driving climate change, we must also prepare for the impacts that are too late to avoid. Across America, states, cities, and communities are taking steps to protect themselves by updating building codes, adjusting the way they manage natural resources, investing in more resilient infrastructure, and planning for rapid recovery from damages that nonetheless occur. The federal government has an important role to play in supporting community-based preparedness and resilience efforts, establishing policies that promote preparedness, protecting critical infrastructure and public resources, supporting science and research germane to preparedness and resilience, and ensuring that federal operations and facilities continue to protect and serve citizens in a changing climate.

The Obama Administration has been working to strengthen America's climate resilience since its earliest days. Shortly after coming into office, President Obama established an Interagency Climate Change Adaptation Task Force and, in October 2009, the President signed an Executive Order directing it to recommend ways federal policies and programs can better prepare the Nation for change. In May 2010, the Task Force hosted the first National Climate Adaptation Summit, convening local and regional stakeholders and decision-makers to identify challenges and opportunities for collaborative action.

In February 2013, federal agencies released Climate Change Adaptation Plans for the first time, outlining strategies to protect their operations, missions, and programs from the effects of climate change. The Department of Transportation, for example, is developing guidance for incorporating climate change and extreme weather event considerations into coastal highway projects, and the Department of Homeland Security is evaluating the challenges of changing conditions in the Arctic and along our Nation's borders. Agencies have also partnered with communities through targeted grant and technical-assistance programs—for example, the Environmental Protection Agency is working with low-lying communities in North Carolina to assess the vulnerability of infrastructure investments to sea level rise and identify solutions to reduce risks. And the Administration has continued, through the U.S. Global Change Research Program, to support science and monitoring to expand our understanding of climate change and its impacts.

Going forward, the Administration will expand these efforts into three major, interrelated initiatives to better prepare America for the impacts of climate change:

I. Building Stronger and Safer Communities and Infrastructure

By necessity, many states, cities, and communities are already planning and preparing for the impacts of climate change. Hospitals must build capacity to serve patients during more frequent heat waves, and urban planners must plan for the severe storms that infrastructure will need to withstand. Promoting on-the-ground planning and resilient infrastructure will be at the core of our work to strengthen America's communities. Specific actions will include:

Directing Agencies to Support Climate-Resilient Investment: The President will direct federal agencies to identify and remove barriers to making climate-resilient investments; identify and remove counterproductive policies that increase vulnerabilities; and encourage and support smarter, more resilient investments, including through agency grants, technical assistance, and other programs, in sectors from transportation and water management to conservation and disaster relief. Agencies will also be directed to ensure that climate risk-management

considerations are fully integrated into federal infrastructure and natural resource management planning. To begin meeting this challenge, the Environmental Protection Agency is committing to integrate considerations of climate change impacts and adaptive measures into major programs, including its Clean Water and Drinking Water State Revolving Funds and grants for brownfields cleanup, and the Department of Housing and Urban Development is already requiring grant recipients in the Hurricane Sandy–affected region to take sea-level rise into account.

Establishing a State, Local, and Tribal Leaders Task Force on Climate Preparedness: To help agencies meet the above directive and to enhance local efforts to protect communities, the President will establish a short-term task force of state, local, and tribal officials to advise on key actions the federal government can take to better support local preparedness and resilience-building efforts. The task force will provide recommendations on removing barriers to resilient investments, modernizing grant and loan programs to better support local efforts, and developing information and tools to better serve communities.

Supporting Communities as they Prepare for Climate Impacts: Federal agencies will continue to provide targeted support and assistance to help communities prepare for climate- change impacts. For example, throughout 2013, the Department of Transportation's Federal Highway Administration is working with 19 state and regional partners and other federal agencies to test approaches for assessing local transportation infrastructure vulnerability to climate change and extreme weather and for improving resilience. The Administration will continue to assist tribal communities on preparedness through the Bureau of Indian Affairs, including through pilot projects and by supporting participation in federal initiatives that assess climate change vulnerabilities and develop regional solutions. Through annual federal agency "Environmental Justice Progress Reports," the Administration will continue to identify innovative ways to help our most vulnerable communities prepare for and recover from the impacts of climate change. The importance of critical infrastructure independence was brought home in the Sandy response. The Federal Emergency Management Agency and the Department of Energy are working with the private sector to address simultaneous restoration of electricity and fuels supply.

Boosting the Resilience of Buildings and Infrastructure: The National Institute of Standards and Technology will convene a panel on disaster-resilience standards to develop a comprehensive, community-based resilience framework and provide guidelines for consistently safe buildings and infrastructure – products that can inform the development of private-sector standards and codes. In addition, building on federal agencies' "Climate Change Adaptation Plans," the Administration will continue efforts to increase the resilience of federal facilities and infrastructure. The Department of Defense, for example, is assessing the relative vulnerability of its coastal facilities to climate change. In addition, the President's FY 2014 Budget proposes $200 million through the Transportation Leadership Awards program for Climate Ready Infrastructure in communities that build enhanced preparedness into their planning efforts, and that have proposed or are ready to break ground on infrastructure projects, including transit and rail, to improve resilience.

Rebuilding and Learning from Hurricane Sandy: In August 2013, President Obama's Hurricane Sandy Rebuilding Task Force will deliver to the President a rebuilding strategy to be implemented in Sandy-affected regions and establishing precedents that can be followed elsewhere. The Task Force and federal agencies are also piloting new ways to support resilience in the Sandy-affected region; the Task Force, for example, is hosting a regional "Rebuilding by Design" competition to generate innovative solutions to enhance resilience. In the transportation sector, the Department of Transportation's Federal Transit Administration (FTA) is dedicating $5.7 billion to four of the area's most impacted transit agencies, of which $1.3 billion will be allocated to locally prioritized projects to make transit systems more resilient to future disasters. FTA will also develop a competitive process for additional funding to identify and support larger, stand-alone resilience projects in the impacted region. To build coastal resilience, the Department of the Interior will launch a $100 million competitive grant program to foster partnerships and promote resilient natural systems while enhancing green spaces and wildlife habitat near urban populations. An additional $250 million will be allocated to support projects for coastal restoration and resilience across the region. Finally, with partners, the U.S. Army Corps of Engineers is conducting a $20 million study to identify strategies to reduce the vulnerability of Sandy-affected coastal communities to future large-scale flood and storm events, and the National Oceanic and Atmospheric Administration will strengthen long-term coastal observations and provide technical assistance to coastal communities.

II. Protecting our Economy and Natural Resources

Climate change is affecting nearly every aspect of our society, from agriculture and tourism to the health and safety of our citizens and natural resources. To help protect critical sectors, while also targeting hazards that cut across sectors and regions, the Administration will mount a set of sector- and hazard-specific efforts to protect our country's vital assets, to include:

Identifying Vulnerabilities of Key Sectors to Climate Change: The Department of Energy will soon release an assessment of climate-change impacts on the energy sector, including power-plant disruptions due to drought and the disruption of fuel supplies during severe storms, as well as potential opportunities to make our energy infrastructure more resilient to these risks. In 2013, the Department of Agriculture and Department of the Interior released several studies outlining the challenges a changing climate poses for America's agricultural enterprise, forests, water supply, wildlife, and public lands. This year and next, federal agencies will report on the impacts of climate change on other key sectors and strategies to address them, with priority efforts focusing on health, transportation, food supplies, oceans, and coastal communities.

Promoting Resilience in the Health Sector: The Department of Health and Human Services will launch an effort to create sustainable and resilient hospitals in the face of climate change. Through a public-private partnership with the healthcare industry, it will identify best practices and provide guidance on affordable measures to ensure that our medical system is resilient to climate impacts. It will also collaborate with partner agencies to share best practices among federal health facilities. And, building on lessons from pilot projects underway in 16 states, it will help train public-health professionals and community leaders to prepare their communities for the health consequences of climate change, including through effective communication of health risks and resilience measures.

Promoting Insurance Leadership for Climate Safety: Recognizing the critical role that the private sector plays in insuring assets and enabling rapid recovery after disasters, the Administration will convene representatives from the insurance industry and other stakeholders to explore best practices for private and public insurers to manage their own processes and investments to account for climate change risks and incentivize policy holders to take steps to reduce their exposure to these risks.

Conserving Land and Water Resources: America's ecosystems are critical to our nation's economy and the lives and health of our citizens. These natural resources can also help ameliorate the impacts of climate change, if they are properly protected.

The Administration has invested significantly in conserving relevant ecosystems, including working with Gulf State partners after the Deepwater Horizon spill to enhance barrier islands and marshes that protect communities from severe storms. The Administration is also implementing climate-adaptation strategies that promote resilience in fish and wildlife populations, forests and other plant communities, freshwater resources, and the ocean. Building on these efforts, the President is also directing federal agencies to identify and evaluate additional approaches to improve our natural defenses against extreme weather, protect biodiversity and conserve natural resources in the face of a changing climate, and manage our public lands and natural systems to store more carbon.

Maintaining Agricultural Sustainability: Building on the existing network of federal climate- science research and action centers, the Department of Agriculture is creating seven new Regional Climate Hubs to deliver tailored, science-based knowledge to farmers, ranchers, and forest landowners. These hubs will work with universities and other partners, including the Department of the Interior and the National Oceanic and Atmospheric Administration, to support climate resilience. Its Natural Resources Conservation Service and the Department of the Interior's Bureau of Reclamation are also providing grants and technical support to agricultural water users for more water-efficient practices in the face of drought and long-term climate change.

Managing Drought: Leveraging the work of the National Disaster Recovery Framework for drought, the Administration will launch a cross-agency National Drought Resilience Partnership as a "front door" for communities seeking help to prepare for future droughts and reduce drought impacts. By linking information (monitoring, forecasts, outlooks, and early warnings) with drought preparedness and longer-term resilience strategies in critical sectors, this effort will help communities manage drought-related risks.

Reducing Wildfire Risks: With tribes, states, and local governments as partners, the Administration has worked to make landscapes more resistant to wildfires, which are exacerbated by heat and drought conditions resulting from climate change.

Federal agencies will expand and prioritize forest and rangeland restoration efforts in order to make natural areas and communities less vulnerable to catastrophic fire. The Department of the Interior and Department of Agriculture, for example, are launching a Western Watershed Enhancement Partnership – a pilot effort in five western states to reduce wildfire risk by removing extra brush and other flammable vegetation around critical areas such as water reservoirs.

Preparing for Future Floods: To ensure that projects funded with taxpayer dollars last as long as intended, federal agencies will update their flood-risk reduction standards for federally funded projects to reflect a consistent approach that accounts for sea-level rise and other factors affecting flood risks.

This effort will incorporate the most recent science on expected rates of sea-level rise (which vary by region) and build on work done by the Hurricane Sandy Rebuilding Task Force, which announced in April 2013 that all federally funded Sandy-related rebuilding projects must meet a consistent flood risk reduction standard that takes into account increased risk from extreme weather events, sea-level rise, and other impacts of climate change.

III. Using Sound Science to Manage Climate Impacts

Scientific data and insights are essential to help government officials, communities, and businesses better understand and manage the risks associated with climate change. The Administration will continue to lead in advancing the science of climate measurement and adaptation and the development of tools for climate-relevant decision-making by focusing on increasing the availability, accessibility, and utility of relevant scientific tools and information. Specific actions will include:

Developing Actionable Climate Science: The President's Fiscal Year 2014 Budget provides more than $2.7 billion, largely through the 13-agency U.S. Global Change Research Program, to increase understanding of climate-change impacts, establish a public-private partnership to explore risk and catastrophe modeling, and develop the information and tools needed by decision-makers to respond to both long-term climate change impacts and near-term effects of extreme weather.

Assessing Climate-Change Impacts in the United States: In the spring of 2014, the Obama Administration will release the third U.S. National Climate Assessment, highlighting new advances in our understanding of climate-change impacts across all regions of the United States and on critical sectors of the economy, including transportation, energy, agriculture, and ecosystems and biodiversity. For the first time, the National Climate Assessment will focus not only on dissemination of scientific information but also on translating scientific insights into practical, useable knowledge that can help decision-makers anticipate and prepare for specific climate-change impacts.

Launching a Climate Data Initiative: Consistent with the President's May 2013 Executive Order on Open Data – and recognizing that freely available open government data can fuel entrepreneurship, innovation, scientific discovery, and public benefits – the Administration is launching a Climate Data Initiative to leverage extensive federal climate-relevant data to stimulate innovation and private-sector entrepreneurship in support of national climate-change preparedness.

Providing a Toolkit for Climate Resilience: Federal agencies will create a virtual climate- resilience toolkit that centralizes access to data-driven resilience tools, services, and best practices, including those developed through the Climate Data Initiative. The toolkit will provide easy access to existing resources as well as new tools, including: interactive sea-level rise maps and a sea-level-rise calculator to aid post-Sandy rebuilding in New York and New Jersey, new NOAA storm surge models and interactive maps from the National Oceanic and Atmospheric Administration that provide risk information by combining tidal data, projected sea levels and storm wave heights, a web-based tool that will allow developers to integrate NASA climate imagery into websites and mobile apps, access to the U.S. Geological Survey's "visualization tool" to assess the amount of carbon absorbed by landscapes, and a Stormwater Calculator and Climate Assessment Tool developed to help local governments assess stormwater-control measures under different precipitation and temperature scenarios.

LEAD INTERNATIONAL EFFORTS TO ADDRESS GLOBAL CLIMATE CHANGE

The Obama Administration is working to build on the actions that it is taking domestically to achieve significant global greenhouse gas emission reductions and enhance climate preparedness through major international initiatives focused on spurring concrete action, including bilateral initiatives with China, India, and other major emitting countries. These initiatives not only serve to support the efforts of the United States and others to achieve our goals for 2020, but also will help us move beyond those and bend the post-2020 global emissions trajectory further. As a key part of this effort, we are also working intensively to forge global responses to climate change through a number of important international negotiations, including the United Nations Framework Convention on Climate Change.

I. Working with Other Countries to Take Action to Address Climate Change

Enhancing Multilateral Engagement with Major Economies: In 2009, President Obama launched the Major Economies Forum on Energy and Climate, a high-level forum that brings together 17 countries that account for approximately 75 percent of global greenhouse gas emissions, in order to support the international climate negotiations and spur cooperative action to combat climate change. The Forum has been successful on both fronts – having contributed significantly to progress in the broader negotiations while also launching the Clean Energy Ministerial to catalyze the development and deployment of clean energy and efficiency solutions. We are proposing that the Forum build on these efforts by launching a major initiative this year focused on further accelerating efficiency gains in the buildings sector, which accounts for approximately one-third of global carbon pollutions from the energy sector.

Expanding Bilateral Cooperation with Major Emerging Economies: From the outset, the Obama Administration has sought to intensify bilateral climate cooperation with key major emerging economies, through initiatives like the U.S.-China Clean Energy Research Center, the U.S.-India Partnership to Advance Clean Energy, and the Strategic Energy Dialogue with Brazil.

We will be building on these successes and finding new areas for cooperation in the second term, and we are already making progress: Just this month, President Obama and President Xi Jinping of China reached an historic agreement at their first summit to work to use the expertise and institutions of the Montreal Protocol to phase down the consumption and production of HFCs, a highly potent greenhouse gas. The impact of phasing out HFCs by 2050 would be equivalent to the elimination of two years' worth of greenhouse gas emissions from all sources.

Combatting Short-Lived Climate Pollutants: Pollutants such as methane, black carbon, and many HFCs are relatively short-lived in the atmosphere, but have more potent greenhouse effects than carbon dioxide. In February 2012, the United States launched the Climate and Clean Air Coalition to Reduce Short-Lived Climate Pollution, which has grown to include more than 30 country partners and other key partners such as the World Bank and the U.N. Environment Programme. Major efforts include reducing methane and black carbon from waste and landfills. We are also leading through the Global Methane Initiative, which works with 42 partner countries and an extensive network of over 1,100 private sector participants to reduce methane emissions.

Reducing Emissions from Deforestation and Forest Degradation: Greenhouse gas emissions from deforestation, agriculture, and other land use constitute approximately one-third of global emissions. In some developing countries, as much as 80 percent of these emissions come from the land sector. To meet this challenge, the Obama Administration is working with partner countries to put in place the systems and institutions necessary to significantly reduce global land-use-related emissions, creating new models for rural development that generate climate benefits, while conserving biodiversity, protecting watersheds, and improving livelihoods.

In 2012 alone, the U.S. Agency for International Development's bilateral and regional forestry programs contributed to reducing more than 140 million tons of carbon dioxide emissions, including through support for multilateral initiatives such as the Forest Investment Program and the Forest Carbon Partnership Facility. In Indonesia, the Millennium Challenge Corporation is funding a five-year "Green Prosperity" program that supports environmentally sustainable, low carbon economic development in select districts.

The Obama Administration is also working to address agriculture-driven deforestation through initiatives such as the Tropical Forest Alliance 2020, which brings together governments, the private sector, and civil society to reduce tropical deforestation related to key agricultural commodities, which we will build upon.

Expanding Clean Energy Use and Cut Energy Waste: Roughly 84 percent of current carbon dioxide emissions are energy-related and about 65 percent of all greenhouse gas emissions can be attributed to energy supply and energy use. The Obama Administration has promoted the expansion of renewable, clean, and efficient energy sources and technologies worldwide through:

- Financing and regulatory support for renewable and clean energy projects
- Actions to promote fuel switching from oil and coal to natural gas or renewables
- Support for the safe and secure use of nuclear power
- Cooperation on clean coal technologies
- Programs to improve and disseminate energy efficient technologies

In the past three years we have reached agreements with more than 20 countries around the world, including Mexico, South Africa, and Indonesia, to support low emission development strategies that help countries to identify the best ways to reduce greenhouse gas emissions while growing their economies. Among the many initiatives that we have launched are:

- The U.S. Africa Clean Energy Finance Initiative, which aligns grant-based assistance with project planning expertise from the U.S. Trade and Development Agency and financing and risk mitigation tools from the U.S. Overseas Private Investment Corporation to unlock up to $1 billion in clean energy financing.
- The U.S.-Asia Pacific Comprehensive Energy Partnership, which has identified $6 billion in U.S. export credit and government financing to promote clean energy development in the Asia-Pacific region.

Looking ahead, we will target these and other resources towards greater penetration of renewables in the global energy mix on both a small and large scale, including through our participation in the Sustainable Energy for All Initiative and accelerating the commercialization of renewable mini-grids. These efforts include:

- *Natural Gas.* Burning natural gas is about one-half as carbon-intensive as coal, which can make it a critical "bridge fuel" for many countries as the world transitions to even cleaner sources of energy. Toward that end, the Obama Administration is partnering with states and private companies to exchange lessons learned with our international partners on responsible development of natural gas resources. We have launched the Unconventional Gas Technical Engagement Program to share best practices on issues such as water management, methane emissions, air quality, permitting, contracting, and pricing to help increase global gas supplies and facilitate development of the associated infrastructure that brings them to market. Going forward, we will promote fuel-switching from coal to gas for electricity production and encourage the development of a global market for gas. Since heavy-duty vehicles are expected to account for 40 percent of increased oil use through 2030, we will encourage the adoption of heavy duty natural gas vehicles as well.
- *Nuclear Power.* The United States will continue to promote the safe and secure use of nuclear power worldwide through a variety of bilateral and multilateral engagements. For example, the U.S. Nuclear Regulatory Commission advises international partners on safety and regulatory best practices, and the Department of Energy works with international partners on research and development, nuclear waste and storage, training, regulations, quality control, and comprehensive fuel leasing options. Going forward, we will expand these efforts to promote nuclear energy generation consistent with maximizing safety and nonproliferation goals.
- *Clean Coal.* The United States works with China, India, and other countries that currently rely heavily on coal for power generation to advance the development and deployment of clean coal technologies. In addition, the U.S. leads the Carbon Sequestration Leadership Forum, which engages 23 other countries and economies on carbon capture and sequestration technologies. Going forward, we will

continue to use these bilateral and multilateral efforts to promote clean coal technologies.

- *Energy Efficiency.* The Obama Administration has aggressively promoted energy efficiency through the Clean Energy Ministerial and key bilateral programs. The cost- effective opportunities are enormous: The Ministerial' s Super-Efficient Equipment and Appliance Deployment Initiative and its Global Superior Energy Performance Partnership are helping to accelerate the global adoption of standards and practices that would cut energy waste equivalent to more than 650 mid-size power plants by 2030. We will work to expand these efforts focusing on several critical areas, including: improving building efficiency, reducing energy consumption at water and wastewater treatment facilities, and expanding global appliance standards.

Negotiating Global Free Trade in Environmental Goods and Services: The U.S. will work with trading partners to launch negotiations at the World Trade Organization towards global free trade in environmental goods, including clean energy technologies such as solar, wind, hydro and geothermal. The U.S. will build on the consensus it recently forged among the 21 Asia-Pacific Economic Cooperation (APEC) economies in this area. In 2011, APEC economies agreed to reduce tariffs to 5 percent or less by 2015 on a negotiated list of 54 environmental goods. The APEC list will serve as a foundation for a global agreement in the WTO, with participating countries expanding the scope by adding products of interest. Over the next year, we will work towards securing participation of countries which account for 90 percent of global trade in environmental goods, representing roughly $481 billion in annual environmental goods trade. We will also work in the Trade in Services Agreement negotiations towards achieving free trade in environmental services.

Phasing Out Subsidies that Encourage Wasteful Consumption of Fossil Fuels: The International Energy Agency estimates that the phase-out of fossil fuel subsidies – which amount to more than $500 billion annually – would lead to a 10 percent reduction in greenhouse gas emissions below business as usual by 2050. At the 2009 G-20 meeting in Pittsburgh, the United States successfully advocated for a commitment to phase out these subsidies, and we have since won similar commitments in other fora such as APEC. President Obama is calling for the elimination of U.S. fossil fuel tax subsidies in his

Fiscal Year (FY) 2014 budget, and we will continue to collaborate with partners around the world toward this goal.

Leading Global Sector Public Financing Towards Cleaner Energy: Under this Administration, the United States has successfully mobilized billions of dollars for clean energy investments in developing countries, helping to accelerate their transition to a green, low-carbon economy. Building on these successes, the President calls for an end to U.S. government support for public financing of new coal plants overseas, except for (a) the most efficient coal technology available in the world's poorest countries in cases where no other economically feasible alternative exists, or (b) facilities deploying carbon capture and sequestration technologies. As part of this new commitment, we will work actively to secure the agreement of other countries and the multilateral development banks to adopt similar policies as soon as possible.

Strengthening Global Resilience to Climate Change: Failing to prepare adequately for the impacts of climate change that can no longer be avoided will put millions of people at risk, jeopardizing important development gains, and increasing the security risks that stem from climate change. That is why the Obama Administration has made historic investments in bolstering the capacity of countries to respond to climate-change risks. Going forward, we will continue to:

- Strengthen government and local community planning and response capacities, such as by increasing water storage and water use efficiency to cope with the increased variability in water supply
- Develop innovative financial risk management tools such as index insurance to help smallholder farmers and pastoralists manage risk associated with changing rainfall patterns and drought
- Distribute drought-resistant seeds and promote management practices that increase farmers' ability to cope with climate impacts.

Mobilizing Climate Finance: International climate finance is an important tool in our efforts to promote low-emissions, climate-resilient development. We have fulfilled our joint developed country commitment from the Copenhagen Accord to provide approximately $30 billion of climate assistance to developing countries over FY 2010-FY 2012. The United States contributed approximately $7.5 billion to this effort over the three year period. Going forward, we will seek to build on this progress as well as focus our efforts on combining our public resources with smart policies to mobilize much larger flows of private investment in low-emissions and climate resilient infrastructure.

II. Leading Efforts to Address Climate Change through International Negotiations

The United States has made historic progress in the international climate negotiations during the past four years. At the Copenhagen Conference of the United Nations Framework Convention on Climate Change (UNFCCC) in 2009, President Obama and other world leaders agreed for the first time that all major countries, whether developed or developing, would implement targets or actions to limit greenhouse emissions, and do so under a new regime of international transparency. And in 2011, at the year-end climate meeting in Durban, we achieved another breakthrough: Countries agreed to negotiate a new agreement by the end of 2015 that would have equal legal force and be applicable to all countries in the period after 2020. This was an important step beyond the previous legal agreement, the Kyoto Protocol, whose core obligations applied to developed countries, not to China, India, Brazil or other emerging countries.

The 2015 climate conference is slated to play a critical role in defining a post-2020 trajectory. We will be seeking an agreement that is ambitious, inclusive and flexible. It needs to be ambitious to meet the scale of the challenge facing us. It needs to be inclusive because there is no way to meet that challenge unless all countries step up and play their part. And it needs to be flexible because there are many differently situated parties with their own needs and imperatives, and those differences will have to be accommodated in smart, practical ways.

At the same time as we work toward this outcome in the UNFCCC context, we are making progress in a variety of other important negotiations as well. At the Montreal Protocol, we are leading efforts in support of an amendment that would phase down HFCs; at the International Maritime Organization, we have agreed to and are now implementing the first-ever sector-wide, internationally applicable energy efficiency standards; and at the International Civil Aviation Organization, we have ambitious aspirational emissions and energy efficiency targets and are working towards agreement to develop a comprehensive global approach.

In: President Obama's Climate Action Plan
Editor: Shane N. Blake
ISBN: 978-1-62808-863-2
© 2013 Nova Science Publishers, Inc.

Chapter 3

PRESIDENTIAL MEMORANDUM: POWER SECTOR CARBON POLLUTION STANDARDS*

Environmental Protection Agency

SUBJECT: Power Sector Carbon Pollution Standards

With every passing day, the urgency of addressing climate change intensifies. I made clear in my State of the Union address that my Administration is committed to reducing carbon pollution that causes climate change, preparing our communities for the consequences of climate change, and speeding the transition to more sustainable sources of energy.

The Environmental Protection Agency (EPA) has already undertaken such action with regard to carbon pollution from the transportation sector, issuing Clean Air Act standards limiting the greenhouse gas emissions of new cars and light trucks through 2025 and heavy duty trucks through 2018. The EPA standards were promulgated in conjunction with the Department of Transportation, which, at the same time, established fuel efficiency standards for cars and trucks as part of a harmonized national program. Both agencies engaged constructively with auto manufacturers, labor unions, States, and other stakeholders, and the resulting standards have received broad support. These standards will reduce the Nation's carbon pollution and dependence on oil, and also lead to greater innovation, economic growth, and cost savings for American families.

---

* This memo, authored by President Barack Obama, was released by The White House, Office of the Press Secretary on June 25, 2013.

The United States now has the opportunity to address carbon pollution from the power sector, which produces nearly 40 percent of such pollution. As a country, we can continue our progress in reducing power plant pollution, thereby improving public health and protecting the environment, while supplying the reliable, affordable power needed for economic growth and advancing cleaner energy technologies, such as efficient natural gas, nuclear power, renewables such as wind and solar energy, and clean coal technology.

Investments in these technologies will also strengthen our economy, as the clean and efficient production and use of electricity will ensure that it remains reliable and affordable for American businesses and families.

By the authority vested in me as President by the Constitution and the laws of the United States of America, and in order to reduce power plant carbon pollution, building on actions already underway in States and the power sector, I hereby direct the following:

Section 1. Flexible Carbon Pollution Standards for Power Plants. (a) Carbon Pollution Standards for Future Power Plants. On April 13, 2012, the EPA published a Notice of Proposed Rulemaking entitled "Standards of Performance for Greenhouse Gas Emissions for New Stationary Sources: Electric Utility Generating Units," 77 Fed. Reg. 22392. In light of the information conveyed in more than two million comments on that proposal and ongoing developments in the industry, you have indicated EPA's intention to issue a new proposal. I therefore direct you to issue a new proposal by no later than September 20, 2013. I further direct you to issue a final rule in a timely fashion after considering all public comments, as appropriate.

(b) *Carbon Pollution Regulation for Modified, Reconstructed, and Existing Power Plants.* To ensure continued progress in reducing harmful carbon pollution, I direct you to use your authority under sections 111(b) and 111(d) of the Clean Air Act to issue standards, regulations, or guidelines, as appropriate, that address carbon pollution from modified, reconstructed, and existing power plants and build on State efforts to move toward a cleaner power sector. In addition, I request that you:

(i) issue proposed carbon pollution standards, regulations, or guidelines, as appropriate, for modified, reconstructed, and existing power plants by no later than June 1, 2014;

(ii) issue final standards, regulations, or guidelines, as appropriate, for modified, reconstructed, and existing power plants by no later than June 1, 2015; and

(iii) include in the guidelines addressing existing power plants a requirement that States submit to EPA the implementation plans required under section 111(d) of the Clean Air Act and its implementing regulations by no later than June 30, 2016.

(c) *Development of Standards, Regulations, or Guidelines for Power Plants*. In developing standards, regulations, or guidelines pursuant to subsection (b) of this section, and consistent with Executive Orders 12866 of September 30, 1993, as amended, and 13563 of January 18, 2011, you shall ensure, to the greatest extent possible, that you:

(i) launch this effort through direct engagement with States, as they will play a central role in establishing and implementing standards for existing power plants, and, at the same time, with leaders in the power sector, labor leaders, non-governmental organizations, other experts, tribal officials, other stakeholders, and members of the public, on issues informing the design of the program;

(ii) consistent with achieving regulatory objectives and taking into account other relevant environmental regulations and policies that affect the power sector, tailor regulations and guidelines to reduce costs;

(iii) develop approaches that allow the use of market-based instruments, performance standards, and other regulatory flexibilities;

(iv) ensure that the standards enable continued reliance on a range of energy sources and technologies;

(v) ensure that the standards are developed and implemented in a manner consistent with the continued provision of reliable and affordable electric power for consumers and businesses; and

(vi) work with the Department of Energy and other Federal and State agencies to promote the reliable and affordable provision of electric power through the continued development and deployment of cleaner technologies and by increasing energy efficiency, including through stronger appliance efficiency standards and other measures.

Sec. 2. General Provisions. (a) This memorandum shall be implemented consistent with applicable law, including international trade obligations, and subject to the availability of appropriations.

(b) Nothing in this memorandum shall be construed to impair or otherwise affect:

(i) the authority granted by law to a department, agency, or the head thereof; or

(ii) the functions of the Director of the Office of Management and Budget relating to budgetary, administrative, or legislative proposals.

(c) This memorandum is not intended to, and does not, create any right or benefit, substantive or procedural, enforceable at law or in equity by any party against the United States, its departments, agencies, or entities, its officers, employees, or agents, or any other person.

(d) You are hereby authorized and directed to publish this memorandum in the *Federal Register*.

BARACK OBAMA

In: President Obama's Climate Action Plan ISBN: 978-1-62808-863-2
Editor: Shane N. Blake © 2013 Nova Science Publishers, Inc.

Chapter 4

EPA STANDARDS FOR GREENHOUSE GAS EMISSIONS FROM POWER PLANTS: MANY QUESTIONS, SOME ANSWERS*

James E. McCarthy

SUMMARY

As President Obama announced initiatives addressing climate change on June 25, 2013, a major focus of attention was on the prospect of greenhouse gas (GHG) emission standards for fossilfueled—mostly coal-fired—electric generating units (EGUs). EGUs (more commonly referred to as power plants) are the largest source of greenhouse gas emissions, accounting for about one-third of total U.S. GHGs. If the country is going to reduce its GHG emissions by significant amounts, as the President has committed to do, emissions from these sources will almost certainly need to be controlled.

The President addressed this issue on June 25 by directing EPA to re-propose GHG emission standards for new EGUs, which the agency had proposed in April 2012, but had not yet finalized. He also directed the agency to develop standards for existing power plants by June 2015.

Under the Clean Air Act, the EPA Administrator has a great deal of flexibility in setting these standards. The statute requires that New Source Performance Standards (NSPS) reflect the degree of emission limitation

* This is an edited, reformatted and augmented version of a Congressional Research Service publication, CRS Report for Congress R43127, prepared for Members and Committees of Congress, from www.crs.gov, dated June 26, 2013.

achievable through application of the best system of emission reduction that has been "adequately demonstrated." The Administrator can take costs, health and environmental impacts, and energy requirements into account in determining what has been adequately demonstrated.

The standard for new EGUs proposed in 2012 would have set a limit of 1,000 pounds of carbon dioxide (CO2) per megawatt-hour of electricity generated—a standard that can be met by new natural gas combined cycle plants without add-on emission controls. Coal-fired plants, however, would find it impossible to meet the standard without controls to capture, compress, and store underground about 45% of the CO2 they produce—a technology referred to as carbon capture and storage (CCS).

Many in the electric power and coal industries view the 2012 proposed standard as effectively prohibiting the construction of new coal-fired power plants. Whether CCS technology has been adequately demonstrated is one question they raise. Other issues involve the cost of compliance, the increased energy required to capture and store carbon, and whether the agency should propose separate standards for gas-fired and coal-fired units. These questions may be addressed in the re-proposal, which the President directed to be issued no later than September 20, 2013.

In its 2012 proposal, EPA maintained that the components of CCS technology have been demonstrated on numerous facilities. Despite this, the agency concluded that no new facilities (other than DOE-sponsored demonstration projects) will actually use CCS in the next 10 years. Given the projected low cost and abundance of natural gas, all new fossil-fueled units are likely to be powered by gas, according to EPA.

Interest in the power plant proposal extends far beyond the electric power and energy industries. Power plants are the first category of sources for which EPA has proposed to require CCS, but the agency expects to propose standards for the GHG emissions of other industries in the next few years. What the agency promulgates in this instance may serve as a precedent.

The potential impacts of the rule also extend beyond new sources, because the agency is obligated under Section 111(d) of the Clean Air Act to promulgate guidelines for existing sources within a category when it promulgates GHG standards for new sources. Using these guidelines, states will be required to develop performance standards for existing sources. These could be less stringent than the NSPS—taking into account, among other factors, the remaining useful life of the existing source—but the standards could have far greater impact than the NSPS, given that they will affect all existing sources.

Many in Congress oppose GHG emission standards. In the 112th Congress, the House passed bills (H.R. 1, H.R. 910, and H.R. 3409) that would have prohibited EPA from promulgating GHG emission standards for any source; the Senate did not follow suit. The President's climate change initiative is likely to stir renewed interest in this subject.

INTRODUCTION

Over the past four years, the Environmental Protection Agency (EPA) has begun to address emissions of greenhouse gases (GHGs)[1] from both mobile and stationary sources, using broad regulatory authority provided by Congress decades ago in the Clean Air Act. Although Congress has never specifically directed EPA to regulate emissions of GHGs, the Clean Air Act as enacted in 1970 and as amended in 1977 and 1990 gave the agency authority to identify air pollutants and promulgate regulations to limit their emission.

From the late 1990s until 2007, EPA and various interested parties debated whether that authority covered greenhouse gases. This debate was settled by the Supreme Court in April 2007, in *Massachusetts v. EPA*. In a 5-4 decision, the Court found that greenhouse gases are unambiguously air pollutants:

> The Clean Air Act's sweeping definition of 'air pollutant' includes 'any air pollution agent or combination of such agents, including any physical, chemical ... substance or matter which is emitted into or otherwise enters the ambient air.... '... Carbon dioxide, methane, nitrous oxide, and hydrofluorocarbons are without a doubt 'physical [and] chemical ... substances[s] which [are] emitted into ... the ambient air.' The statute is unambiguous.[2]

Since the Court's *Massachusetts* decision, EPA has addressed GHG emissions in a number of steps, among them:

- In December 2009, the agency laid the groundwork for regulations by finding that emissions of greenhouse gases endanger public health and welfare, and that GHGs from motor vehicles cause or contribute to that endangerment.
- In May 2010, the agency promulgated GHG emission standards for model year 2012-2016 cars and light trucks.
- In January 2011, the agency began requiring permits and the imposition of Best Available Control Technology on new stationary

sources (and major modifications of existing sources) that emit more than a threshold amount of GHGs.

- In September 2011, EPA promulgated GHG emission standards for model year 2014-2018 medium- and heavy-duty trucks.
- In October 2012, the agency promulgated a second phase of GHG emission standards for cars and light trucks, covering model years 2017-2025.[3]

As extensive as these actions may seem, they have had relatively minor impacts on GHG emissions to date. The rules are prospective, and in most cases have not yet taken effect.

The auto and truck manufacturing industries have been the major focus of the GHG regulations; in both cases, they are eager to improve fuel economy (coincidentally reducing GHG emissions), because the high cost of fuel has affected consumer purchasing decisions over the last five years. Almost without exception, the major companies in these industries have supported EPA's GHG standards.

The stationary source permitting requirements have yet to affect most sources. EPA and the states issued fewer than 50 GHG permits to stationary sources in the year following the requirement's implementation, because the emission threshold for requiring permits was set at a high level, and because few new facilities were being constructed in the recession's aftermath.[4]

Ultimately, if EPA is to reduce the nation's GHG emissions, as the President has committed to do, it will have to issue emission standards for broad categories of existing stationary sources. EPA took the first step toward setting such standards on April 13, 2012, with the proposal of standards for new electric generating units (EGUs). In a June 25, 2013, memorandum to the EPA Administrator, the President directed the agency to re-propose those standards by September 20, 2013, finalize them "in a timely fashion after considering all public comments," and develop guidelines for existing EGUs by June 1, 2015.[5]

As shown in Table 1, EGUs—principally, coal-fired power plants—are the most significant U.S. source of greenhouse gases, accounting for about one-third of the nation's total emissions. With the principal mobile source categories already subject to GHG regulations, EPA will have addressed the sources of more than half of all U.S. emissions once it promulgates regulations for existing EGUs.

NEW SOURCE PERFORMANCE STANDARDS (NSPS)

To control GHG emissions from stationary sources, EPA intends to use Section 111 of the Clean Air Act, which requires the agency to set New Source Performance Standards (NSPS). As noted, EPA proposed the first such NSPS/GHG standard, for electric generating units, on April 13, 2012.[6] The agency has also committed to the promulgation of NSPS for petroleum refineries, although it is unclear when those standards will be proposed.[7]

Under Section 111, the EPA Administrator is required to set standards for categories of new (or substantially modified) major sources if, in his judgment, they cause or contribute significantly to air pollution which may reasonably be anticipated to endanger public health or welfare.[8] Over the past four decades, EPA Administrators have used this authority to set emission standards for numerous sources of conventional pollutants, such as sulfur dioxide or nitrogen oxides. The standards are to reflect the degree of emission limitation achievable through application of the best "adequately demonstrated" system of emission reduction.

Table 1. U.S. Greenhouse Gas Emissions, 2010, by Source Category (in million tons of CO_2 equivalent)

Source	2010 Emissions	% of Total U.S. GHG Emissions
Electricity Generation	2306	34%
- Coal-fired	1840	27%
- Natural gas fired	405	6%
- Oil-fired	31	<1%
Transportation	1834	27%
- Passenger cars	769	11%
- Light-duty trucks	320	5%
- Medium and heavy duty trucks	390	6%
- Aircraft[a]	144	2%
Industry[b]	1394	20%
Agriculture	495	7%
Commercial	382	6%
Residential	365	5%

Source: U.S. EPA, Inventory of U.S. Greenhouse Gas Emissions and Sinks: 1990-2010, April 15, 2012.

[a] Excludes international use of aviation fuel, which would add another 1% (72 million tons).

[b] Numerous industries, including iron and steel production, petroleum refining, cement kilns, and others.

The Administrator can take costs, health impacts, environmental impacts, and energy requirements into account in setting the standards; he can distinguish among classes, types, and sizes of sources; and he must review the standards at least every eight years.

Regulating Existing Sources Under Section 111

In addition to standards for new sources, Section 111 requires that EPA develop guidelines applicable to GHG emissions from *existing* units whenever it promulgates standards for new sources in a category (Section 111(d)). Using the guidelines, states would be required to develop performance standards for existing sources. These standards could be less stringent than the NSPS, taking into account, among other factors, the remaining useful life of the existing source to which the standard applies. Nevertheless, these standards might have far greater impact than the NSPS, given that existing power plants are the largest U.S. source of GHG emissions.

The authority to control existing sources is particularly important in sectors like the electric utility industry, where old units can continue operating for decades. The average coal-fired power plant in the United States is more than 40 years old. Without the authority to control emissions from such existing facilities, it could be decades before emissions from most power plants would be controlled.

How quickly Section 111(d) standards will be applied to existing sources has been an open question, however. EPA must first propose and promulgate guidelines, following which the states will be given time to develop implementation plans. In the President's June 25 memorandum, he requested that EPA:

i. issue proposed guidelines for modified, reconstructed, and existing power plants by no later than June 1, 2014;
ii. issue final guidelines by no later than June 1, 2015; and
iii. include in the guidelines addressing existing power plants a requirement that states submit to EPA the implementation plans required under section 111(d) of the Clean Air Act by no later than June 30, 2016.

Following approval of the plans, the act envisions case-by-case determinations of emission limits. Thus, it is likely to be several years before existing power plants or other stationary sources are subject to emission limits for GHGs.

EPA'S NSPS PROPOSAL FOR EGU GREENHOUSE GAS EMISSIONS

As noted, the President directed EPA in his June 25 memorandum to re-propose the NSPS standards for power plant carbon pollution. To understand what issues the agency may address in the re-proposed standards, this report discusses the 2012 proposal and several issues raised during its consideration.

On April 13, 2012, the Environmental Protection Agency proposed NSPS for greenhouse gas emissions from electric generating units.[9] EPA called the proposal the Carbon Pollution Standard for New Power Plants. By statute, the rule was to have been finalized within a year, but the agency did not do so. One reason is that the public comment period generated more than 2 million comments, which must be considered in the development of a final rule.

Emission Limits

The proposed 2012 standard would have set a limit of 1,000 pounds of carbon dioxide (CO_2) per megawatt-hour (MWh) of electricity generated. This standard can be met by new natural gas combined cycle plants without add-on emission controls. Coal-fired plants, however, would find it impossible to meet the standard without controls to capture and store some of the CO_2 they produce. EPA estimates that a supercritical pulverized coal-fired power plant without such controls produces roughly 1,800 lbs CO_2 /MWh of electricity, so a plant subject to the standard would need to reduce emissions by about 45%.

Carbon capture and storage (CCS) technology that might be used to reduce CO_2 emissions is the subject of much recent research and demonstration.[10] It poses a number of challenges, not the least of which is the additional energy it consumes. The energy required to run equipment that can remove CO_2 from an emission stream (referred to as the "parasitic load") is currently in the range of 30% on most demonstration projects. In addition, a CCS-equipped unit would incur costs for underground storage of the captured CO_2 and possibly significant costs for building and operating a pipeline to transport the CO_2 to the storage location.

To address the concerns of those who maintain that CCS technology is not yet available, or who expect the technology to improve (bringing down costs) as research and demonstration continue, the agency has proposed an alternative under which coal-fired facilities would be allowed to average their emissions over a 30-year period: during the first 10 years, such facilities could emit up to 1,800 lbs CO_2 /MWh; the facility would then need to reduce emissions to 600 lbs/MWwh for the following 20 years.

EPA also solicited comment on whether the emissions standard that reflects CCS should be somewhat higher or lower than 1,000 lb CO_2/MWh, and whether the flexible (first 10 years) emissions standard that reflects supercritical efficiency should be somewhat higher or lower than 1,800 lbs CO_2/MWh.[11]

Exemptions from the Proposed Standard

NSPS are different from other Clean Air Act emission standards in that, once a standard is final, its effective date is retroactive to the date when a *proposed* standard was published in the *Federal Register*, rather than after the date of the final standard's promulgation. Under a strict reading of this requirement, a final standard would apply to any EGU on which construction began after April 13, 2012.

EPA's 2012 proposal for GHG emissions from new power plants would have exempted most facilities from this requirement, however. The agency would exempt:

- new units that had permits and started construction within 12 months of the proposal (i.e., by April 13, 2013);
- units looking to renew permits that are part of a Department of Energy demonstration project, provided that these units started construction within 12 months of the proposal;
- new units located in non-continental areas, which include Hawaii and the territories;
- existing units, including modifications, such as changes needed to meet other air pollution standards;
- simple cycle natural gas units, which are generally designed to produce peaking power;

- new units that do not burn fossil fuels (e.g., units that burn biomass only).

EPA referred to the first two groups (i.e., units that have permits and began construction by April 13, 2013), as "transitional" units, and says there are about 15 of them.[12] They include units that are participating in DOE CCS funding programs. Whether the proposed exemptions will be modified in the re-proposed rule is unclear.

EPA's Cost-Benefit Analysis

EPA's Regulatory Impact Analysis (RIA) for the 2012 proposal concluded that "even in the absence of this rule, existing and anticipated economic conditions in the marketplace will lead electricity generators to choose technologies that meet the proposed standards."[13] Those conditions include the abundance and low projected cost of natural gas, the many state requirements that increasing amounts of electricity come from renewable sources, and the increasing cost of coal-fired electricity due to higher coal prices and new emission standards for emissions of conventional and toxic air pollutants. These factors combined make it likely that almost all new generation will come from natural gas combined cycle or renewable sources, according to EPA. Using the current version of the Integrated Planning Model, a model developed by ICF Inc. that EPA and many industry sources have used to analyze the impacts of regulations since the 1980s, EPA ran several different scenarios to analyze the sensitivity of the results to various assumptions. These included a high electricity demand scenario, a low gas recovery scenario (which results in higher prices for natural gas), and a scenario combining both assumptions. None of these assumptions caused the model to indicate construction of new coal-fired capacity.[14] The analyses found that "the price of natural gas would have to increase to approximately $10/mmBtu [million Btu] for coal boilers without CCS to become competitive with combined cycle natural gas units, which is projected to be very unlikely."[15] The current price of natural gas, on June 26, 2013, was about $3.75/mmBtu.[16]

With no new coal plants (except for those identified as exempted from the rule, above), EPA saw little quantifiable impact from its prospective NSPS promulgation. As the RIA states, "... EPA anticipates that the proposed EGU GHG NSPS will result in negligible CO_2 emission changes, energy impacts, quantified benefits, costs, and economic impacts by 2020."[17]

The RIA argued the case for the rule as a backstop, in case its projections of market conditions are inaccurate:

> This NSPS provides legal assurance that any new coal-fired plants must limit CO2 emissions. Rather than relying solely on changeable energy market conditions to provide low emissions from new power plants in the future, this rule prevents the possible construction of uncontrolled, high-emitting new sources that might continue to emit at high levels for decades, contributing to accumulation of CO_2 in the atmosphere.... In addition, EPA intends this rule to send a clear signal about the future of CCS technology that, in conjunction with other policies such as Department of Energy (DOE) financial assistance, the agency estimates will support development and demonstration of CCS technology from coal-fired plants at commercial scale.... [18]

Thus, the agency maintained that the proposed rule would assist in the deployment of CCS technology, noting that regulatory uncertainty may have hindered its deployment. The proposed rule's preamble noted, for example,

> American Electric Power (AEP)'s recent deferral of a large-scale CCS retrofit demonstration project on one of its coal-fired power plants because the State's utility regulators would not approve CCS without a regulatory requirement to reduce CO2. The standard established in this proposal would help create the regulatory certainty that CCS is the path forward for new coal-fired generation.[19]

In addition, the promulgation of New Source Performance Standards, even if the standards have little or no effect on those new sources, serves as the precondition for standards affecting existing units. The latter are described in Section 111(d) of the Clean Air Act as "standards of performance for any existing source ... to which a standard of performance under this section would apply if such existing source were a new source."

KEY QUESTIONS REGARDING THE 2012 PROPOSED RULE

Many in the electric power and coal industries view the standard proposed in 2012, if finalized, as effectively prohibiting the construction of new coal-fired power plants other than those granted exemptions. Whether carbon capture and storage (CCS) technology has been adequately demonstrated is one question they raise. Other questions involve whether the cost of compliance, assuming CCS is available, and the increased energy required to capture and store carbon should lead the Administrator to propose a less stringent standard. Whether the agency should have proposed separate standards for gas-fired and coal-fired units, and whether the proposed standard is barred by statutory language that prohibits the Administrator from requiring the installation and operation of any particular emission reduction system are other issues. These questions and perhaps others may be addressed when EPA re-proposes the GHG standard pursuant to the President's directive.

Has CCS Been Adequately Demonstrated?

EPA maintains that the components of CCS technology have been demonstrated on numerous facilities. According to the agency, "... at present, CCS is technologically feasible for implementation at new coal-fired power plants and its core components (CO_2 capture, compression, transportation, and storage) have already been implemented at commercial scale."[20] Specifically, in its *Federal Register* notice proposing the standard, the agency said:

- Capture of CO_2 from industrial gas streams has occurred since the 1930s using a variety of approaches to separate CO_2 from other gases.
- Carbon dioxide has been transported via pipelines in the U.S. for nearly 40 years. Approximately 50 million metric tons of CO_2 are transported each year through 3,600 miles of pipelines. Moreover, a review of the 500 largest CO_2 point sources in the United States shows that 95% are within 50 miles of a possible geologic sequestration site, which would lower transportation costs.
- With respect to carbon sequestration/storage, there are at least four commercial integrated CCS facilities sequestering CO_2 into deep geologic formations and applying a suite of technologies to monitor and verify that the CO_2 remains sequestered.[21]

Critics of the agency maintain that even if the components have been demonstrated, there is no plant that captures and stores CO_2 on the scale of a large coal-fired power plant. There are several large power plants currently under development that will demonstrate CCS at commercial scale when completed, but none of these is currently operational, and several planned projects have been abandoned for a variety of reasons.[22]

Should Cost and/or Energy Considerations Have Led EPA to Propose a Less Stringent Standard?

Although it maintains that CCS has been adequately demonstrated, EPA concedes, based on DOE estimates, that "using today's commercially available CCS technologies would add around 80 percent to the cost of electricity for a new pulverized coal (PC) plant, and around 35 percent for a new advanced gasification-based (IGCC) plant."[23] The Congressional Budget Office, in a June 2012 report, reached essentially the same conclusion.[24] Much of this increased cost results from what is termed a "parasitic" energy load: capturing CO_2, compressing it, transporting it, and injecting it underground would use as much as 30% of the electricity that a coal-fired plant produces.

Both EPA and the Congressional Budget Office, among others, assume that the cost and energy penalty can be reduced through research, development, and demonstration, and both view EPA regulation as one of the policy tools that could lead to reduced cost by forcing the development of better technology. Experience suggests that such "learning by doing" will lower the cost, but the road to what might be a competitive technology is a long one, and given the availability of other power sources (such as natural gas, renewables, and nuclear) with lower or no carbon emissions, it is not clear that the electric power industry will be motivated to pursue it.

Despite the potentially high cost of currently available CCS technology, the agency "does not anticipate this rule will have any impacts on the price of electricity, employment or labor markets, or the US economy."[25] Other than "transitional" coal-fired units exempt from the proposed standard and demonstration projects supported by DOE or other incentives, no new coal-fired units will actually use CCS in the next 10 years: given the low cost and projected abundance of natural gas, all new fossil-fueled units are likely to be powered by gas, according to EPA.

> Forecasts suggesting that new coal is unlikely to be built by 2020 have been shown to be robust under a range of alternative assumptions that influence the industry's decisions to build new power plants. For example, EIA typically supplements the AEO [Annual Energy Outlook] with a series of distinct scenarios that explore specific issues and examine

a future state of the world that deviates from the core parameter estimates that underlie the AEO reference case. Even under alternative scenarios where assumptions might improve the relative economic value of building new coal-fired power plants, the AEO 2011 does not project new coal capacity being built through 2025, beyond the coal capacity already planned outside of the modeling. Relevant scenarios include higher economic growth forecast, lower cost of coal supply, lower capital costs of fossil fuel-fired energy technologies, and less optimistic natural gas supply.[26]

Should EPA Have Proposed Separate Standards for Coal- and Gas-Fired Units?

New Source Performance Standards promulgated previously for conventional pollutants (such as sulfur dioxide) emitted by electric generating units have generally distinguished coal- and gas-fired units. As EPA notes in the preamble to the proposed standard, in setting standards for conventional pollutants or air toxics, it was not appropriate to combine coal-fired and gas-fired units in a single category, because "although coal-fired EGUs have an array of control options for criteria and air toxic air pollutants to choose from, those controls generally do not reduce their ... emissions to the level of conventional emissions from natural gas-fired EGUs."[27]

Critics of the GHG proposal have taken EPA's statement a step further, stating frequently that combining coal-fired and gas-fired units in a single category is "unprecedented."[28] This is not actually the case: in 1998, EPA promulgated NSPS for emissions of nitrogen oxides (NOx) that imposed a single emission standard on all fossil-fueled EGUs.[29] In *Lignite Energy Council v. U. S. EPA*,[30] the D.C. Circuit Court of Appeals squarely addressed the argument that a single fuel-neutral standard was impermissible and rejected it.

Is the Standard Barred by Statutory Language?

Critics of the proposed standard maintain that in setting the standard at 1,000 lbs CO_2/MWh, EPA would effectively require coal-fired power plants to add CCS to any new unit. They maintain that such a requirement violates Section 111(b)(5) of the Clean Air Act, which states that unless he determines that it is not feasible to prescribe or enforce a standard of performance, the EPA Administrator is not authorized to require a new source "to install and operate any particular technological system of continuous emission reduction to comply with any new source standard of performance."

Whether a court would find that EPA is imposing a "particular technological system" might come down to an interpretation of the term. There are several different technologies for carbon capture under development. Does the fact that they all result in capturing CO_2 make them a "particular technological system"? Or would a court find that the option of switching fuels to lower emissions means that sources can comply with the standard without having to install and operate a particular technological system?[31]

GUIDELINES FOR EXISTING POWER PLANTS

The potential impacts of the NSPS rule extend beyond new sources, because the agency is obligated under Section 111(d) of the act to promulgate guidelines for *existing* sources within a category whenever it promulgates GHG standards for *new* sources.[32] Using these guidelines, states will be required to develop performance standards for existing sources. These could be less stringent than the NSPS—taking into account, among other factors, the remaining useful life of the existing source to which the standard applies—but the standards could have far greater impact than the NSPS, given that existing plants account for one-third of total U.S. GHG emissions.

The average coal-fired power plant is about 40 years old; some are more than 60 years old. The older plants are generally less efficient than newer units, and most operate only a small percentage of the time. Thus, the agency might choose to set a guideline based on a less costly approach than application of CCS to the units' emissions. In presentation slides that the agency has used in stakeholder discussions, emphasis has been placed on improving efficiency as a preferred approach to reducing GHG emissions.[33]

In recent months, considerable attention has been given to a proposal by the Natural Resources Defense Council (NRDC) as to how Section 111(d) guidelines might be structured.[34] Under NRDC's plan, each state would be given an emission budget based on the mix of fuels used by EGUs in the state to generate electricity in a base period (2008-2010 in the NRDC proposal). States with more coal-fired generation would receive higher budgets than those with more natural gas or renewable sources. The state budgets would be reduced in phases, with a target reduction of about 26% in GHG emissions overall by 2020, compared to 2005 emission levels.

EGUs could comply in a variety of ways: by shifting power dispatch to lower emitting plants (and thus running higher emitting plants less often), by switching fuels, by co-firing lower emitting fuels with coal, by retiring their least efficient plants, by efficiency improvements at existing plants, or by reducing demand. EGUs could average, bank, or trade emission credits; as a result, individual units would have an emissions target, but they could exceed the target if they had sufficient credits obtained from earlier reductions or from other units in the state's electric system. Since the goal would be to reduce emissions overall, rather than in specific states, states might also combine their markets for allowances, giving individual electric generating units and companies additional flexibility.

Whatever their form, a key question regarding the 111(d) guidelines has been when EPA would propose them. The President has now resolved this question, as noted earlier, requesting the agency to propose the guidelines by June 1, 2014, finalize them by June 1, 2015, and require the states to submit implementation plans by June 30, 2016. In cases where a state fails to submit a satisfactory plan, Section 111(d) also gives EPA the authority to prescribe and enforce a federal plan.

CONGRESSIONAL RESPONSES

Many in Congress oppose EPA standards for GHG emissions. The House passed two bills in 2011 (H.R. 1 and H.R. 910) that would have prohibited EPA from promulgating GHG emission standards for any source, and it repeated itself in September 2012 with H.R. 3409, the Stop the War on Coal Act. The Senate did not follow suit.

Legislation to limit or prevent EPA regulatory action is considered possible in the 113th Congress, and may be given a boost by the President's proposal to use the Clean Air Act's Section 111 authority.

Enacting such legislation faces hurdles similar to those encountered in the last Congress, however. Although the House could take action to block NSPS regulations, the Senate is less likely to do so.[35] If the House and Senate did act to limit Executive Branch authority, a bill sent to the President would almost certainly be subject to a veto, given the President's recent statements regarding the importance of dealing with climate issues.[36]

Thus, the most likely role for Congress in addressing NSPS standards for GHGs would appear to be one of oversight.

End Notes

[1] Six greenhouse gases, or groups of gases, are addressed by EPA regulatory actions: carbon dioxide (CO2), methane (CH4), nitrous oxide (N2O), sulfur hexafluoride (SF6), hydrofluorocarbons (HFCs), and perfluorocarbons (PFCs). Of these, carbon dioxide, produced by combustion of fossil fuels, is by far the most prevalent, accounting for 85% of annual emissions of the combined group when measured as CO2 equivalents.

[2] Massachusetts v. EPA, 549 U.S. 497 (2007). For additional discussion, see CRS Report RS22665, The Supreme Court's Climate Change Decision: Massachusetts v. EPA.

3 For a more complete listing of the actions EPA and other agencies have taken in the wake of the Court's Massachusetts decision, see CRS Report R41103, Federal Agency Actions Following the Supreme Court's Climate Change Decision in Massachusetts v. EPA: A Chronology, by Robert Meltz.

[4] At a House Energy and Commerce hearing on June 29, 2012, EPA Assistant Administrator Gina McCarthy stated that EPA and the states had issued 44 permits for greenhouse gas emissions. Previously, in a March 2012 Federal Register notice, the agency stated that EPA and state permitting authorities had issued 18 permits and had received an additional 50 permit applications. See U.S. EPA, "Prevention of Significant Deterioration and Title Five Greenhouse Gas Tailoring Rule Step 3, GHG Plantwide Applicability Limitations and GHG Synthetic Minor Limitations," 77 Federal Register 14233, March 8, 2012.

[5] Office of the Press Secretary, The White House, "Power Sector Carbon Pollution Standards," Memorandum for the Administrator of the Environmental Protection Agency, June 25, 2013, at http://www.whitehouse.gov/the-press-office/2013/06/25/presidential-memorandum -power-sector-carbon-pollution-standards.

[6] The proposed standard, and additional background material, are available on EPA's website at http://www.epa.gov/ carbonpollutionstandard/actions.html.

[7] On December 23, 2010, EPA announced that it was settling a lawsuit filed by 11 states, two municipalities, and three environmental groups over its 2008 decision not to establish New Source Performance Standards for GHG emissions from petroleum refineries. According to the agency, refineries are the second-largest direct stationary source of GHGs in the United States and there are cost-effective strategies for reducing these emissions. The agency agreed to propose NSPS for new refinery facilities and emissions guidelines for existing facilities by December 10, 2011, and to make a final decision on the proposed actions by November 10, 2012. As of this writing (June 2013), the standards and guidelines had not been proposed.

[8] The language is similar to the endangerment and cause-or-contribute findings EPA promulgated for motor vehicles on December 15, 2009 ("Endangerment and Cause or Contribute Findings for Greenhouse Gases Under Section 202(a) of the Clean Air Act," 74 Federal Register 66496).

[9] U.S. EPA, Standards of Performance for Greenhouse Gas Emissions for New Stationary Sources: Electric Utility Generating Units, Proposed Rule, 77 Federal Register 22392, April 13, 2012. A link to the proposed standards and additional background material can be found at http://www.epa.gov/carbonpollutionstandard/actions.html.

[10] CRS has multiple reports on CCS technology. For an overview, see CRS Report R42532, Carbon Capture and Sequestration (CCS): A Primer, by Peter Folger. For a discussion of available policy tools to encourage the development of CCS, see CRS Report R41325, Carbon Capture: A Technology Assessment, by Peter Folger. For a discussion of the Department of Energy's efforts to develop CCS technology, see CRS Report R42496, Carbon Capture and Sequestration: Research, Development, and Demonstration at the U.S. Department of Energy, by Peter Folger, and CRS Report R43028, FutureGen: A Brief History and Issues for Congress, by Peter Folger.

[11] For additional information, see 77 Federal Register 22406, April 13, 2012.

[12] 77 Federal Register 22395, April 13, 2012.

[13] U.S. EPA, Regulatory Impact Analysis for the Proposed Standards of Performance for Greenhouse Gas Emissions for New Stationary Sources: Electric Utility Generating Units, March 2012, p. ES-3, at http://www.epa.gov/ carbonpollutionstandard/pdfs/20120327 proposalRIA.pdf. Hereafter "EPA RIA."

[14] EPA RIA, pp. 5-10 through 5-14.

[15] EPA RIA, p. 5-1. The RIA adds, elsewhere, the following caveats: "It is important to note that this analysis is based on assumptions regarding the average national cost of generation at new facilities. As reported by the EIA [DOE's Energy Information Administration], there is expected to be significant spatial variation in the costs of new generation due to design differences, labor wage and productivity differences, location adjustments, among other potential differences. EPA acknowledges that there is some uncertainty around these estimates, and is unable to provide estimates for all variants. However, the results are expected to hold for the majority of situations. The analysis also does not explicitly consider new units designed to combust waste coal or petroleum coke (pet coke), which may be affected by this rule, but also may exhibit different local economics." (footnotes omitted) See EPA RIA, p. 5-17.

[16] Bloomberg, Energy and Oil Prices, at http://www.bloomberg.com/energy/. On p. 5-17, EPA's RIA specifies that coal-fired power without CCS becomes competitive when natural gas reaches $9.60 per million Btu. EPA notes: "To put this gas price point into historical context, $9.60/MMBtu is higher than any average annual gas price (in 2007 dollars) observed over the last 10 years, and it has only been reached temporarily in 8 of the last 120 months." EPA RIA, p. 5-33.

[17] EPA RIA, p. ES-3.

[18] EPA RIA, pp. ES-3 and ES-4.

[19] 77 Federal Register 22396-2397, April 13, 2012 (footnote omitted).

[20] 77 Federal Register 22414, April 13, 2012.

[21] Ibid., p. 22415 (footnotes omitted).

[22] For details on the demonstration projects, see CRS Report R42496, Carbon Capture and Sequestration: Research, Development, and Demonstration at the U.S. Department of Energy, by Peter Folger, and Congressional Budget Office, Federal Efforts to Reduce the Cost of Capturing and Storing Carbon Dioxide, June 2012. One of the uncertainties faced by CCS is how to address liability concerns associated with CCS technology. For a discussion of this issue, see CRS Report RL34307, Legal Issues Associated with the Development of Carbon Dioxide Sequestration Technology, by Adam Vann and Paul W. Parfomak.

[23] 77 Federal Register 22415-22416, April 13, 2012.

[24] Congressional Budget Office, Federal Efforts to Reduce the Cost of Capturing and Storing Carbon Dioxide, June 2012, pp. 7-9.

[25] EPA RIA, p. ES-3.

[26] EPA RIA, pp. 5-10 and 5-11.

[27] 77 Federal Register 22411, April 13, 2012.

[28] See, for example, "Single Carbon Standard for Power Plants Breaks EPA Precedent, Commenters Say," Daily Environment Report, June 28, 2012.

[29] U.S. EPA, "Revision of Standards of Performance for Nitrogen Oxide Emissions from New Fossil-Fuel Fired Steam Generating Units; Revisions to Reporting Requirements for Standards of Performance for New Fossil-Fuel Fired Steam Generating Units," 63 Federal Register 49442, September 16, 1998. Discussion of that rule's fuel neutral approach begins on page 49445.

[30] 198 F.3d 930, 933 (D.C. Cir. 1999).

[31] The situation faced by an EGU under the proposed standard is somewhat analogous to that of an EGU needing to reduce sulfur dioxide emissions. In that case, too, a unit may switch fuels or it may use a specific type of technology (flue gas desulfurization, generally referred to as a "scrubber").

[32] This statutory requirement was addressed in the December 23, 2010, settlement agreement between EPA and 11 states, 2 municipalities, and 3 environmental groups. In the agreement, EPA stated that "it would be appropriate for it to concurrently propose performance standards for GHG emissions from new and modified EGUs under CAA section 111(b) ... and emission guidelines for GHG emissions from existing affected EGUs pursuant to CAA section 111(d)...." The agency agreed to propose and take final action on a "rule under Section 111(d) that includes emission guidelines for GHGs from existing EGUs" on the same schedule as that set for the NSPS standards.

[33] See, for example, U.S. EPA, Office of Air Quality Planning and Standards, "Rulemaking for Greenhouse Gas Emissions from Electric Utility Steam Generating Units," Presentation Slides, May, 2011, at http://www.epa.gov/air/ tribal/pdfs/presentation-ghggas emissions utility05-25-2011.pdf.

[34] Natural Resources Defense Council, Closing the Power Plant Carbon Pollution Loophole, December 2012, 87 pages, at http://www.nrdc.org/air/pollution-standards/files/pollution-standards-report.pdf.

[35] The Senate could consider a resolution of disapproval under the Congressional Review Act, which can be brought to the floor under rules that prohibit a filibuster, thus requiring a simple majority vote. For discussion of that option, see CRS Report R41212, EPA Regulation of Greenhouse Gases: Congressional Responses and Options.

[36] The President spoke of the importance of addressing climate issues on the night of his re-election in 2012, in his second Inaugural Address, and in the 2013 State of the Union speech, in addition to the June 25, 2013 climate speech.

In: President Obama's Climate Action Plan ISBN: 978-1-62808-863-2
Editor: Shane N. Blake © 2013 Nova Science Publishers, Inc.

Chapter 5

CARBON CAPTURE AND SEQUESTRATION (CCS): A PRIMER*

Peter Folger

SUMMARY

Carbon capture and sequestration (or storage)—known as CCS—has attracted congressional interest as a measure for mitigating global climate change because large amounts of carbon dioxide (CO_2) emitted from fossil fuel use in the United States are potentially available to be captured and stored underground and prevented from reaching the atmosphere.

Large, industrial sources of CO_2, such as electricity-generating plants, are likely initial candidates for CCS because they are predominantly stationary, single-point sources.

Electricity generation contributes over 40% of U.S. CO_2 emissions from fossil fuels. Currently, U.S. power plants do not capture large volumes of CO_2 for CCS.

Several projects in the United States and abroad—typically associated with oil and gas production—are successfully capturing, injecting, and storing CO_2 underground, albeit at relatively small scales. The oil and gas industry in the United States injects nearly 50 million tons of CO_2 underground each year for the purpose of enhanced oil recovery (EOR). The volume of CO_2 envisioned for CCS as a climate

---

* This is an edited, reformatted and augmented version of a Congressional Research Service publication, CRS Report for Congress R42532, from www.crs.gov, prepared for Members and Committees of Congress, dated May 14, 2012.

mitigation option is overwhelming compared to the amount of CO_2 used for EOR.

According to the U.S. Department of Energy (DOE), the United States has the potential to store billions of tons of CO_2 underground and keep the gas trapped there indefinitely. Capturing and storing the equivalent of decades or even centuries of CO_2 emissions from power plants (at current levels of emissions) suggests that CCS has the potential to reduce U.S. greenhouse gas emissions substantially while allowing the continued use of fossil fuels.

An integrated CCS system would include three main steps: (1) capturing and separating CO_2 from other gases; (2) purifying, compressing, and transporting the captured CO_2 to the sequestration site; and (3) injecting the CO_2 in subsurface geological reservoirs or storing it in the oceans. Deploying CCS technology on a commercial scale would be a vast undertaking.

The CCS process, although simple in concept, would require significant investments of capital and of time.

Capital investment would be required for the technology to capture CO_2 and for the pipeline network to transport the captured CO_2 to the disposal site.

Time would be required to assess the potential CO_2 storage reservoir, inject the captured CO_2, and monitor the injected plume to ensure against leaks to the atmosphere or to underground sources of drinking water, potentially for years or decades until injection activities cease and the injected plume stabilizes.

Three main types of geological formations in the United States are being considered for storing large amounts of CO_2: oil and gas reservoirs, deep saline reservoirs, and unmineable coal seams. The deep ocean also has a huge potential to store carbon; however, direct injection of CO_2 into the deep ocean is controversial, and environmental concerns have forestalled planned experiments in the open ocean. Mineral carbonation—reacting minerals with a stream of concentrated CO_2 to form a solid carbonate—is well understood, but it is still an experimental process for storing large quantities of CO_2.

Large-scale CCS injection experiments are only beginning in the United States to test how different types of reservoirs perform during CO_2 injection of a million tons of CO_2 or more. Results from the experiments will undoubtedly be crucial to future permitting and site approval regulations.

Acceptance by the general public of large-scale deployment of CCS may be a significant challenge. Some of the large-scale injection tests could garner information about public acceptance, as citizens become familiar with the concept, process, and results of CO_2 injection tests in their local communities.

INTRODUCTION

Carbon capture and sequestration (or storage)—known as CCS—is a physical process that involves capturing manmade carbon dioxide (CO_2) at its source and storing it before its release to the atmosphere. CCS could reduce the amount of CO_2 emitted to the atmosphere despite the continued use of fossil fuels. An integrated CCS system would include three main steps:

(1) capturing CO_2 and separating it from other gases; (2) purifying, compressing, and transporting the captured CO_2 to the sequestration site; and (3) injecting the CO_2 in subsurface geological reservoirs or storing it in the oceans. As a measure for mitigating global climate change, CCS has attracted congressional interest and support because several projects in the United States and abroad—typically associated with oil and gas production—are successfully capturing, injecting, and storing CO_2 underground, albeit at relatively small scales. The oil and gas industry in the United States injects approximately 48 million metric tons of CO_2 underground each year to help recover oil and gas resources (a process known as enhanced oil recovery, or EOR).[1] Potentially, much larger amounts of CO_2 produced from electricity generation—approximately 2.2 billion metric tons per year, over 40% of the total CO_2 emitted in the United States from fossil fuels (see *Table 1*)—could be targeted for large-scale CCS.

Table 1. Sources for CO_2 Emissions in the United States from Combustion of Fossil Fuels

Sources	CO_2 Emissions[a] (millions of metric tons)	Percent of Total
Electricity generation	2,216.8	42%
Transportation	1,745.5	33%
Industrial	777.8	15%
Residential	340.2	6%
Commercial	224.2	4%
Total	5,304.5	100%

Source: U.S. Environmental Protection Agency (EPA), Inventory of U.S. Greenhouse Emissions and Sinks: 1990- 2010, Table ES-3 (2012); see http://epa.gov/climatechange/emissions/usinventoryreport.html.

[a] CO_2 emissions in millions of metric tons for 2010; excludes emissions from U.S. territories.

Fuel combustion accounts for 94.4% of all U.S. CO_2 emissions.[2] Electricity generation contributes the largest proportion of CO_2 emissions compared to other types of fossil fuel use in the United States (*Table 1*). Electricity-generating plants are among the most likely initial candidates for capture, separation, and storage or reuse of CO_2 because they are predominantly large, stationary, single-point sources of emissions. Large industrial facilities, such as cement-manufacturing, ethanol, or hydrogen production plants, that produce large quantities of CO_2 as part of the industrial process are also good candidates for CO_2 capture and storage.[3]

This report is a brief summary of what CCS is, how it is supposed to work, why it has gained the interest and support of some members of Congress, and what some of the challenges are to its implementation and deployment across the United States.

This report covers only CCS and not other types of carbon sequestration activities whereby CO_2 is removed from the atmosphere and stored in vegetation or soils, such as forests and agricultural lands.[4]

CO_2 CAPTURE

The first step in CCS is to capture CO_2 at the source and produce a concentrated stream for transport and storage. Currently, three main approaches are available to capture CO_2 from large-scale industrial facilities or power plants: (1) post-combustion capture, (2) pre-combustion capture, and (3) oxy-fuel combustion capture. For power plants, current commercial CO_2 capture systems could operate at 85%-95% capture efficiency.[5] The capture phase of the CCS process, however, may be 80% or more of the total costs for CCS.[6]

Post-Combustion Capture

This process involves extracting CO_2 from the flue gas following combustion of fossil fuels or biomass. Several commercially available technologies, some involving absorption using chemical solvents, can in principle be used to capture large quantities of CO_2 from flue gases. U.S. commercial electricity-generating plants currently do not capture large volumes of CO_2 because they are not required to and there are no economic incentives to do so. Nevertheless, the post-combustion capture process includes proven technologies that are commercially available today.

Pre-Combustion Capture

This process separates CO_2 from the fuel by combining the fuel with air and/or steam to produce hydrogen for combustion and a separate CO_2 stream that could be stored. The most common technologies today use steam reforming, in which steam is employed to extract hydrogen from natural gas.[7]

Oxy-Fuel Combustion Capture

This process uses oxygen instead of air for combustion and produces a flue gas that is mostly CO_2 and water, which are easily separable, after which the CO_2 can be compressed, transported, and stored. The U.S. Department of Energy's (DOE) flagship CCS demonstration project, FutureGen, plans to retrofit an existing power unit with an oxy-fuel combustion unit.[8]

CO_2 TRANSPORT

Pipelines are the most common method for transporting CO_2 in the United States. Currently, approximately 4,100 miles of pipeline transport CO_2 in the United States, predominately to oil and gas fields, where it is used for EOR.[9] Transporting CO_2 in pipelines is similar to transporting petroleum products like natural gas and oil; it requires attention to design, monitoring for leaks, and protection against overpressure, especially in populated areas.[10]

Using ships may be feasible when CO_2 must be transported over large distances or overseas. Ships transport CO_2 today, but at a small scale because of limited demand. Liquefied natural gas, propane, and butane are routinely shipped by marine tankers on a large scale worldwide. Rail cars and trucks can also transport CO_2, but this mode would probably be uneconomical for large-scale CCS operations.

Costs for pipeline transport vary, depending on construction, operation and maintenance, and other factors, including right-of-way costs, regulatory fees, and more. The quantity and distance transported will mostly determine costs, which will also depend on whether the pipeline is onshore or offshore, the level of congestion along the route, and whether mountains, large rivers, or frozen ground are encountered. Shipping costs are unknown in any detail, however, because no large-scale CO_2 transport system (in millions of metric tons of CO_2 per year, for example) is operating. Ship costs might be lower than pipeline transport for distances greater than 1,000 kilometers and for less than a few million metric tons of CO_2 ($MtCO_2$)[11] transported per year.[12]

Even though regional CO_2 pipeline networks currently operate in the United States for enhanced EOR, developing a more expansive network for CCS could pose numerous regulatory and economic challenges. Some of these include questions about pipeline network requirements, economic regulation, utility cost recovery, regulatory classification of CO_2 itself, and pipeline safety.[13]

CO_2 SEQUESTRATION

Three main types of geological formations are being considered for carbon sequestration: (1) depleted oil and gas reservoirs, (2) deep saline reservoirs, and (3) unmineable coal seams. In each case, CO_2 would be injected in a supercritical state—a relatively dense liquid—below ground into a porous rock formation that holds or previously held fluids. When CO_2 is injected at depths greater than 800 meters in a typical reservoir, the pressure keeps the injected CO_2 in a supercritical state (dense like a liquid, fluid like a gas) and thus it is less likely to migrate out of the geological formation. Injecting CO_2 into deep geological formations uses existing technologies that have been primarily developed and used by the oil and gas industry, and that could potentially be adapted for long-term storage and monitoring of CO_2. Other underground injection applications in practice today, such as natural gas storage, deep injection of liquid wastes, and subsurface disposal of oil-field brines, can also provide valuable experience and information for sequestering CO_2 in geological formations.[14]

The storage capacity for CO_2 storage in geological formations is potentially huge if all the sedimentary basins in the world are considered (see discussion below of storage capacity estimates for the United States).[15] The suitability of any particular site, however, depends on many factors, including proximity to CO_2 sources and other reservoir-specific qualities like porosity, permeability, and potential for leakage. For CCS to succeed, it is assumed that each reservoir type would permanently store the vast majority of injected CO_2, keeping the gas isolated from the atmosphere in perpetuity.

Oil and Gas Reservoirs

Pumping CO_2 into oil and gas reservoirs to boost production (EOR) is practiced in the petroleum industry today. The United States is a world leader in this technology, and oil and gas operators inject approximately 48 $MtCO_2$ underground each year to help recover oil and gas resources.[16] Most of the CO_2 used for EOR in the United States comes from naturally occurring geologic formations, however, not from industrial sources. Using CO_2 from industrial emitters has appeal because the costs of capture and transport from the facility could be partially offset by revenues from oil and gas production.

Carbon dioxide can be used for EOR onshore or offshore. To date, most CO_2 projects associated with EOR are onshore, with the bulk of U.S. activities in west Texas. Carbon dioxide can also be injected into oil and gas reservoirs that are completely depleted, which would serve the purpose of long-term sequestration, but without any offsetting benefit from oil and gas production.

Advantages and Disadvantages

Depleted or abandoned oil and gas fields, especially in the United States, are considered prime candidates for CO_2 storage for several reasons:

- oil and gas originally trapped did not escape for millions of years, demonstrating the structural integrity of the reservoir;
- extensive studies for oil and gas typically have characterized the geology of the reservoir;
- computer models have often been developed to understand how hydrocarbons move in the reservoir, and the models could be applied to predicting how CO_2 could move; and
- infrastructure and wells from oil and gas extraction may be in place and might be used for handling CO_2 storage.

Some of these features could also be disadvantages to CO_2 sequestration. Wells that penetrate from the surface to the reservoir could be conduits for CO_2 release if they are not plugged properly. Care must be taken not to overpressure the reservoir during CO_2 injection, which could fracture the caprock—the part of the formation that formed a seal to trap oil and gas—and subsequently allow CO_2 to escape. Also, shallow oil and gas fields (those less than 800 meters deep, for example) may be unsuitable because CO_2 may form a gas instead of a denser liquid and could escape to the surface more easily. In addition, oil and gas fields that are suitable for EOR may not necessarily be located near industrial sources of CO_2. Costs to construct pipelines to connect sources of CO_2 with oil and gas fields may, in part, determine whether an EOR operation using industrial sources of CO_2 is feasible.

Although the United States injects nearly 50 $MtCO_2$ underground each year for the purposes of EOR, that amount represents approximately 2% of the CO_2 emitted from fossil fuel electricity generation alone. The sheer volume of CO_2 envisioned for CCS as a climate mitigation option is overwhelming compared to the amount of CO_2 used for EOR. It may be that EOR will increase in the future, depending on economic, regulatory, and technical factors, and more CO_2 will be sequestered as a consequence. It is also likely that EOR would only account for a small fraction of the total amount of CO_2 injected underground in the future, even if CCS becomes a significant component in an overall scheme to substantially reduce CO_2 emissions to the atmosphere.

The In Salah and Weyburn Projects

The In Salah Project in Algeria is the world's first large-scale effort to store CO_2 in a natural gas reservoir.[17] At In Salah, CO_2 is separated from the produced natural gas (the gas contains approximately 5.5% CO_2) and then reinjected into the same formation. Approximately 17 $MtCO_2$ are planned to be captured and stored over the lifetime of the project at a rate of slightly more than 1 Mt per year.[18]

The Weyburn Project in south-central Canada uses CO_2 produced from a coal gasification plant in North Dakota for EOR, injecting up to 5,000 tCO_2 per day into the formation and recovering oil.[19] Approximately 20 $MtCO_2$ are expected to remain in the formation over the lifetime of the project.[20]

Deep Saline Reservoirs

Some rocks in sedimentary basins contain saline fluids—brines or brackish water unsuitable for agriculture or drinking. As with oil and gas, deep saline reservoirs can be found onshore and offshore; in fact, they are often part of oil and gas reservoirs and share many characteristics. The oil industry routinely injects brines recovered during oil production into saline reservoirs for disposal.[21] Using suitably deep saline reservoirs for CO_2 sequestration has advantages: (1) they are more widespread in the United States than oil and gas reservoirs and thus have greater probability of being close to large point sources of CO_2; and (2) saline reservoirs have potentially the largest reservoir capacity of the three types of geologic formations.

Advantages and Disadvantages

Although deep saline reservoirs potentially have huge capacity to store CO_2, estimates of lower and upper capacities vary greatly, reflecting a higher degree of uncertainty in how to measure storage capacity.[22] Actual storage capacity may have to be determined on a case-by-case basis. Estimates of storage capacity for the United States from the DOE Regional Sequestration Partnership Program are discussed below.

From estimates of the potential storage capacity in saline reservoirs, it is likely that the vast majority of CO_2 injected underground would be stored in these formations, assuming that CCS were deployed on a commercial scale across the United States. In addition to their potential capacity, deep saline reservoirs underlie large portions of the country, and could be more easily accessible to large, stationary sources of CO_2 than oil and gas reservoirs or coal seams. *Figure 1* shows broad outlines of sedimentary basins containing the deep saline reservoirs, and the locations of a variety of stationary sources of CO_2.

The DOE Regional Sequestration Partnership Program has conducted simulations, field studies, small-scale injection projects, and is now beginning a phase of large-scale injection demonstration projects to investigate the suitability of deep saline reservoirs.[23] Because of the potentially vast amounts of CO_2 that could be sequestered, these experiments could shed light on the potential for leakage of CO_2 from the reservoir, and test the ability to detect the movement of CO_2 underground as well as to detect leaks through overlying cap rocks.

In addition to the possibility of CO_2 leakage, injection of millions of tons of CO_2 will displace large volumes of brine in the deep saline reservoirs. One disadvantage is therefore the possibility that displaced brine could leak into underground sources of drinking water. Ultimately, CO_2 will likely dissolve into the brine, but that could take decades.

Also, injecting large volumes of fluid into the subsurface has the potential to trigger earthquakes, especially if the CO_2 is injected into an undetected fault. Presumably, evaluating the potential storage site prior to beginning injection will limit the potential for triggering earthquakes (also referred to as "induced seismicity"), but there is no guarantee that fluid could not migrate to faulted or fractured rocks over the course of many years and induce an earthquake.

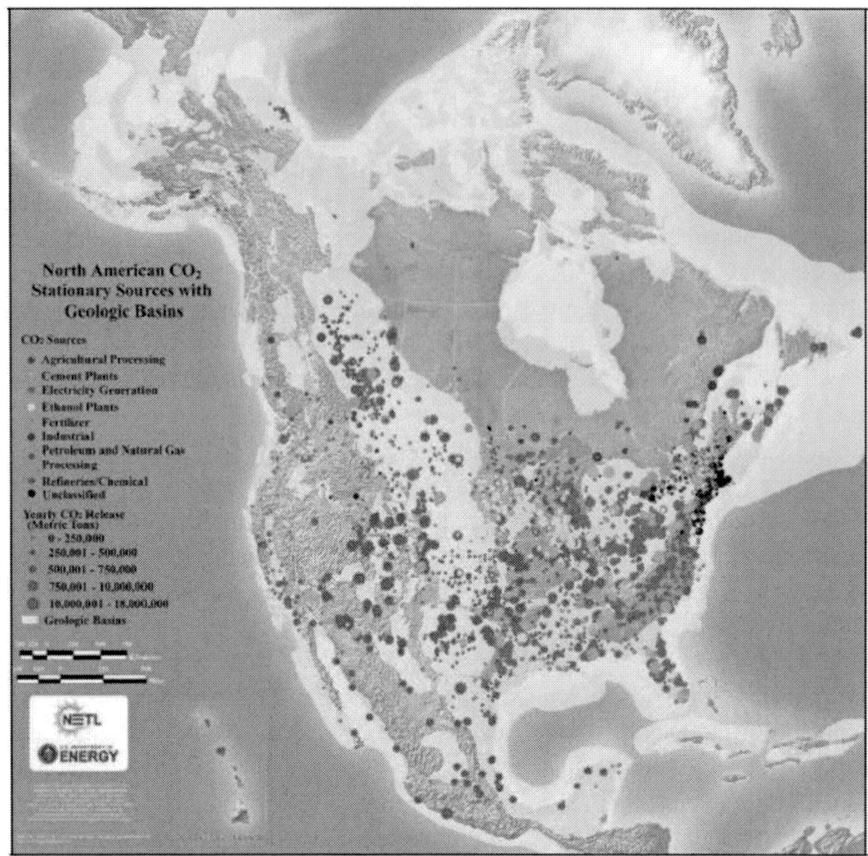

Source: U.S. Department of Energy, National Energy Technology Laboratory, "2010 Carbon Sequestration Atlas of the United States and Canada, Third Edition," http://www.netl.doe.gov/technologies/carbon_seq/refshelf/ atlasIII/index.html.

Note: Not all geologic basins have deep saline reservoirs suitable for carbon sequestration.

Figure 1. Stationary Sources of CO2 in North America and Underlying Geologic Basins.

The issue of induced seismicity has recently been linked to the injection and disposal of produced waters from oil and gas fields.[24]

Unlike sequestration in existing oil and gas fields, injecting into deep saline reservoirs may take place in regions of the country that have not experienced drilling activities. Public opposition may arise to activities on the surface—such as land clearing, building of new roads, transport of heavy equipment, and operation of drill rigs—but also to the concept of disposing

CO_2 underground near residences and communities. There is at least one example of public opposition to CO_2 injection leading to cancellation of a project in Europe.[25]

The Sleipner Project

The Sleipner Project in the North Sea is the first commercial-scale operation for sequestering CO_2 in a deep saline reservoir. The Sleipner project has been operating since 1996, and it injects and stores approximately 2,800 tCO_2 per day, or about 1 $MtCO_2$ per year.[26] Carbon dioxide is separated from natural gas production at the nearby Sleipner West Gas Field, compressed, and then injected 800 meters below the seabed of the North Sea into the Utsira formation, a sandstone reservoir 200-250 meters (650-820 feet) thick containing saline fluids. Monitoring has indicated the CO_2 has not leaked from the saline reservoir, and computer simulations suggest that the CO_2 will eventually dissolve into the saline water, reducing the potential for leakage in the future.

Another CO_2 sequestration project, similar to Sleipner, began in the Barents Sea in April 2008 (the Snohvit Project),[27] and is injecting approximately 2,000 tCO_2 per day below the seafloor. A larger project is being planned in western Australia (the Gorgon Project)[28] that would inject 9,000 tCO_2 per day when at full capacity. Similar to the Sleipner and Snohvit operations, the Gorgon plans to strip CO_2 from produced natural gas and inject it into deep saline formations for permanent storage.

Unmineable Coal Seams

U.S. coal resources not mineable with current technology are those where the coal beds[29] are not thick enough, or are too deep, or whose structural integrity is inadequate for mining. Even if they cannot be mined, coal beds are commonly permeable and can trap gases, such as methane, which can be extracted (a resource known as coal-bed methane, or CBM). Methane and other gases are physically bound (adsorbed) to the coal. Studies indicate that CO_2 binds even more tightly to coal than methane.[30] Carbon dioxide injected into permeable coal seams could displace methane, which could be recovered by wells and brought to the surface, providing a source of revenue to offset the costs of CO_2 injection.

Advantages and Disadvantages

Unmineable coal seam injection projects would need to assess several factors in addition to the potential for CBM extraction. These include depth, permeability, coal bed geometry (a few thick seams, not several thin seams), lateral continuity and vertical isolation (less potential for upward leakage), and other considerations. Once CO_2 is injected into a coal seam, it would likely remain there unless the seam is depressurized or the coal is mined. Many unmineable coal seams in the United States are located relatively near electricity-generating facilities, which could reduce the distance and cost of transporting CO_2 from large point sources to storage sites.

Not all types of coal beds are suitable for CBM extraction. Without the coal-bed methane resource, the sequestration process would be less economically attractive. However, the displaced methane would need to be combusted or captured because methane itself is a more potent greenhouse gas than CO_2. Once burned, methane produces mostly CO_2 and water.

Without ongoing commercial experience, storing CO_2 in coal seams has significant uncertainties compared to the other two types of geological storage discussed. According to IPCC, unmineable coal seams have the smallest potential capacity for storing CO_2 globally compared to oil and gas fields or deep saline formations. The latest assessment from DOE also indicates that unmineable coal seams in the United States have less potential capacity than U.S. oil and gas fields for storing CO_2. (See following discussion.) No commercial CO_2 injection and sequestration projects in coal beds are currently underway in the United States.

GEOLOGICAL STORAGE CAPACITY FOR CO_2 IN THE UNITED STATES

As *Figure 1* indicates, geologic basins containing at least one of each of these three types of potential CO_2 reservoirs occur across most of the United States, in relative proximity to many large point sources of CO_2, such as fossil fuel power plants or cement plants. The DOE Regional Sequestration Partnership Program has produced estimates of the potential storage capacity for each of these types of reservoirs and published the estimates in a Carbon Sequestration Atlas. The 2010 Carbon Sequestration Atlas (third edition) updates the 2008 version (second edition), and a summary of the storage estimates for both editions is compared in *Table 2*.[31]

The Carbon Sequestration Atlas was compiled from estimates of geological storage capacity made by seven separate regional partnerships (government-industry collaborations fostered by DOE) that each produced estimates for different regions of the United States and parts of Canada.

According to DOE, geographical differences in fossil fuel use and sequestration potential across the country led to a regional approach to assessing CO_2 sequestration potential.[32] The Carbon Sequestration Atlas reflects some of the regional differences; for example, not all of the regional partnerships identified unmineable coal seams as potential CO_2 reservoirs. Other partnerships identified geological formations unique to their regions—such as organic-rich shales in the Illinois Basin, or flood basalts in the Columbia River Plateau—as other types of possible reservoirs for CO_2 storage.

Table 2 indicates a lower and upper range for sequestration potential in deep saline formations and for unmineable coal seams, but only a single estimate for oil and gas fields. Comparison between the 2008 and 2010 estimates indicates small changes between the two estimates for oil and gas fields, but relatively larger changes in estimates for deep saline formations and unmineable coal seams. It is clear from the table that DOE considers estimates for oil and gas fields much better constrained than for the other types of reservoirs. The amount and types of data from oil and gas fields, such as production history, and reservoir volume calculations, often represent decades of experience in the oil and gas industry. In the Carbon Sequestration Atlas, oil and gas reservoirs were assessed at the field level (i.e., on a finer scale and in more detail) than deep saline formations or unmineable coal seams, which were assessed at the basin level (i.e., at a coarser scale and in less detail).

Other methodologies for and estimates of the geological sequestration potential have been released or are underway. For example, the Energy Independence and Security Act of 2007 (EISA, P.L. 110-140) directed the Department of the Interior (DOI) to develop a single methodology for an assessment of the national potential for geologic storage of carbon dioxide. The U.S. Geological Survey (USGS) released an initial methodology in 2009. In response to external comments and reviews, the USGS revised its initial methodology in a 2010 report. [33] According to DOE, the USGS effort will allow refinement of the estimates provided in the 2008 Carbon Sequestration Atlas, and will incorporate uncertainty in the capacity estimates.[34] In addition, DOE notes that its methodology will incorporate results from large-scale carbon sequestration demonstration projects now underway, and that it will update its CO_2 storage estimates every two years. The DOE Sequestration

Atlas should probably be considered an evolving assessment of U.S. reservoir capacity for CO_2 storage.[35]

The total lower estimate (sum of the three reservoir types) from the 2010 Carbon Sequestration Atlas shown in *Table 2* indicates the potential to store the equivalent of 830 years of CO_2 emissions from electricity generation in the United States at current emission rates (2.2 billion tons per year). The total upper estimate indicates the potential for over 9,000 years of CO_2 emissions from electricity generation.

Table 2. Geological Sequestration Potential for the United States and Parts of Canada (billion metric tons of CO_2)

Reservoir type	Lower estimate (2010)	Lower estimate (2008)	% Change	Upper estimate (2010)	Upper estimate (2008)	% Change
Oil and gas fields	143	138	+3.6%	143	138	+3.6%
Deep saline formations	1,653	3,297	-50%	20,213	12,618	+60%
Unmineable coal seams	60	157	-62%	117	178	-34%
Totals	1,856	3,592	-48%	20,473	12,934	+58%

Source: 2008 and 2010 Carbon Sequestration Atlases.

DEEP OCEAN SEQUESTRATION

The world's oceans contain approximately 50 times the amount of carbon stored in the atmosphere and nearly 10 times the amount stored in plants and soils.[36] The oceans today take up—act as a net sink for—approximately 1.7 $GtCO_2$ per year. About 45% of the CO_2 released from fossil fuel combustion and land use activities during the 1990s has remained in the atmosphere, while the remainder has been taken up by the oceans, vegetation, or soils on the land surface.[37] Without the ocean sink, atmospheric CO_2 concentration would be increasing more rapidly. Ultimately, the oceans could store more than 90% of all the carbon released to the atmosphere by human activities, but the process takes thousands of years.[38] The ocean's capacity to absorb atmospheric CO_2 may change, however, and possibly even decrease in the future.[39] Also, studies indicate that as more CO_2 enters the ocean from the atmosphere, the surface waters are becoming more acidic.[40]

Advantages and Disadvantages

Although the surface of the ocean is becoming more concentrated with CO_2, the surface waters and the deep ocean waters generally mix very slowly, on the order of decades to centuries. Injecting CO_2 directly into the deep ocean would take advantage of the slow rate of mixing, allowing the injected CO_2 to remain sequestered until the surface and deep waters mix and CO_2 concentrations equilibrate with the atmosphere. What happens to the CO_2 would depend on how it is released into the ocean, the depth of injection, and the temperature of the seawater.

Carbon dioxide injected at depths shallower than 500 meters typically would be released as a gas, and would rise towards the surface. Most of it would dissolve into seawater if the injected CO_2 gas bubbles were small enough.[41] At depths below 500 meters, CO_2 can exist as a liquid in the ocean, although it is less dense than seawater. After injection below 500 meters, CO_2 would also rise, but an estimated 90% would dissolve in the first 200 meters. Below 3,000 meters in depth, CO_2 is a liquid and is denser than seawater; the injected CO_2 would sink and dissolve in the water column or possibly form a CO_2 pool or lake on the sea bottom. Some researchers have proposed injecting CO_2 into the ocean bottom sediments below depths of 3,000 meters, and immobilizing the CO_2 as a dense liquid or solid CO_2 hydrate.[42] Deep storage in ocean bottom sediments, below 3,000 meters in depth, might potentially sequester CO_2 for thousands of years.[43]

The potential for ocean storage of captured CO_2 is huge, but environmental impacts on marine ecosystems and other issues may determine whether large quantities of captured CO_2 will ultimately be stored in the oceans. Also, deep ocean storage is in a research stage, and the effects of scaling up from small research experiments, using less than 100 liters of CO_2,[44] to injecting several $GtCO_2$ into the deep ocean are unknown.

Injecting CO_2 into the deep ocean would change ocean chemistry, locally at first, and assuming that hundreds of $GtCO_2$ were injected, would eventually produce measurable changes over the entire ocean.[45] The most significant and immediate effect would be the lowering of pH, increasing the acidity of the water. A lower pH may harm some ocean organisms, depending on the magnitude of the pH change and the type of organism. Actual impacts of deep sea CO_2 sequestration are largely unknown, however, because scientists know very little about deep ocean ecosystems.[46]

Environmental concerns led to the cancellation of the largest planned experiment to test the feasibility of ocean sequestration in 2002. A scientific consortium had planned to inject 60 tCO_2 into water over 800 meters deep near the Kona coast on the island of Hawaii. Environmental organizations opposed the experiment on the grounds that it would acidify Hawaii's fishing grounds, and that it would divert attention from reducing greenhouse gas emissions.[47] A similar but smaller project with plans to release more than 5 tCO_2 into the deep ocean off the coast of Norway, also in 2002, was cancelled by the Norway Ministry of the Environment after opposition from environmental groups.[48]

Sequestering Under the Seabed

Deep ocean sequestration, as discussed here, is different from injecting CO_2 beneath the seabed into depleted oil and gas reservoirs or deep saline formations.

The Sleipner project discussed above is an example of injection beneath the seafloor, but not injection into the ocean waters. Sequestering CO_2 under the seabed on the U.S. continental shelf would eliminate the need to negotiate with local landowners over the rights to surface land and to the pore space in the subsurface.

However, it would also require developing an offshore infrastructure to transport and inject the captured CO_2, along with all the other challenges of evaluating the potential offshore reservoir, including monitoring the injected CO_2, and providing for liability and ownership of the CO_2 after injection has ceased.

MINERAL CARBONATION

Another option for sequestering CO_2 produced by fossil fuel combustion involves converting CO_2 to solid inorganic carbonates, such as $CaCO_3$ (limestone), using chemical reactions. When this process occurs naturally, it is known as "weathering" and takes place over thousands or millions of years. The process can be accelerated by reacting a high concentration of CO_2 with minerals found in large quantities on the Earth's surface, such as olivine or serpentine.[49] Mineral carbonation has the advantage of sequestering carbon in solid, stable minerals that can be stored without risk of releasing carbon to the atmosphere over geologic time scales.[50]

Mineral carbonation involves three major activities: (1) preparing the reactant minerals—mining, crushing, and milling—and transporting them to a processing plant, (2) reacting the concentrated CO_2 stream with the prepared minerals, and (3) separating the carbonate products and storing them in a suitable repository.

Advantages and Disadvantages

Mineral carbonation is well understood and can be applied at small scales, but is at an early phase of development as a technique for sequestering large amounts of captured CO_2. Large volumes of silicate oxide minerals are needed, from 1.6 to 3.7 metric tons of silicates per tCO_2 sequestered. Thus, a large-scale mineral carbonation process needs a large mining operation to provide the reactant minerals in sufficient quantity.[51] Large volumes of solid material would also be produced, between 2.6 and 4.7 metric tons of materials per tCO_2 sequestered, or 50%-100% more material to be disposed of by volume than originally mined. Because mineral carbonation is in the research and experimental stage, estimating the amount of CO_2 that could be sequestered by this technique is difficult.

One possible type of geological reservoir for CO_2 storage—major flood basalts[52] such as those on the Columbia River Plateau—is being explored for its potential to react with CO_2 and form solid carbonates in situ (in place). Instead of mining, crushing, and milling the reactant minerals, as discussed above, CO_2 would be injected directly into the basalt formations and would react with the rock over time and at depth to form solid carbonate minerals. Large and thick formations of flood basalts occur globally, and many have characteristics—such as high porosity and permeability—that are favorable to storing CO_2. Those characteristics, combined with the tendency of basalt to react with CO_2, could result in a large-scale conversion of the gas into stable, solid minerals that would remain underground for geologic time. The DOE regional carbon sequestration partnerships are exploring the possibility of using Columbia River Plateau flood basalts in the Pacific Northwest for storing CO_2.[53]

CURRENT ISSUES AND FUTURE CHALLENGES

A primary goal of developing and deploying CCS is to allow large industrial facilities, such as fossil fuel power plants and cement plants, to operate while reducing their CO_2 emissions by 80%-90%. Such reductions would presumably reduce the likelihood of continued climate warming from greenhouse gases by slowing the rise in atmospheric concentrations of CO_2. To achieve the overarching goal of reducing the likelihood of continued climate warming would depend, in part, on how fast and how widely CCS could be deployed throughout the economy.

The additional cost of installing CCS on CO_2-emitting facilities is a primary challenge to the adoption and deployment of CCS in the United States.

Major increases in CO_2 capture technology efficiency will likely produce the greatest relative cost savings for CCS systems, but challenges also face the transport and storage components of CCS. Ideally, storage reservoirs for CO_2 would be located close to sources, obviating the need to build a large pipeline infrastructure to deliver captured CO_2 for underground sequestration. If CCS moves to widespread implementation, however, some areas of the country may not have adequate reservoir capacity nearby, and may need to construct pipelines from sources to reservoirs. Identifying and validating sequestration sites would need to account for CO_2 pipeline costs, for example, if the economics of the sites are to be fully understood. If this is the case, there would be questions to be resolved regarding pipeline network requirements, economic regulation, utility cost recovery, regulatory classification of CO_2 itself, and pipeline safety. In addition, Congress may be called upon to address federal jurisdictional authority over CO_2 pipelines under existing law, and whether additional legislation may be necessary if a CO_2 pipeline network grows and crosses state lines.

Although DOE has identified substantial potential storage capacity for CO_2, particularly in deep saline formations, large-scale injection experiments are only beginning in the United States to test how different types of reservoirs perform during CO_2 injection. Data from the experiments will undoubtedly be crucial to future permitting and site approval regulations.

In addition, liability, ownership, and long-term stewardship for CO_2 sequestered underground are issues that would need to be resolved before CCS is deployed commercially.

Some states are moving ahead with state-level geological sequestration regulations for CO_2, so federal efforts to resolve these issues at a national level would likely involve negotiations with the states. Acceptance by the general public of large-scale deployment of CCS may be a significant challenge if the majority of CCS projects involve private land. Some of the large-scale injection tests could garner information about public acceptance, as local communities become familiar with the concept, process, and results of CO_2 injection tests. Apart from the question of how the public would accept the likely higher cost for electricity generated from plants with CCS, how a growing CCS infrastructure of pipelines, injection wells, underground reservoirs, and other facilities would be accepted by the public is as yet unknown.

End Notes

[1] U.S. Department of Energy, National Energy Technology Laboratory, Carbon Sequestration Through Enhanced Oil Recovery, (March, 2008), at http://www.netl.doe.gov/publications/ factsheets/program/Prog053.pdf.

[2] U.S. Environmental Protection Agency (EPA), Inventory of U.S. Greenhouse Emissions and Sinks: 1990-2010, p. ES-7. The percentage refers to U.S. emissions in 2010; see http:// epa.gov/climatechange/emissions/usinventoryreport.html.

[3] Intergovernmental Panel on Climate Change (IPCC) Special Report: Carbon Dioxide Capture and Storage, 2005. (Hereafter referred to as IPCC Special Report.)

[4] For more information about carbon sequestration in forests and agricultural lands, see CRS Report RL31432, Carbon Sequestration in Forests, by Ross W. Gorte; CRS Report RL33898, Climate Change: The Role of the U.S. Agriculture Sector, by Renée Johnson; and CRS Report R40186, Biochar: Examination of an Emerging Concept to Sequester Carbon, by Kelsi Bracmort. For more information about carbon exchanges between the oceans, atmosphere, and land surface, see CRS Report RL34059, The Carbon Cycle: Implications for Climate Change and Congress, by Peter Folger.

[5] IPCC Special Report, p. 107.

[6] See, for example, John Deutch et al., The Future of Coal, Massachusetts Institute of Technology, An Interdisciplinary MIT Study, 2007, Executive Summary, p. xi.

[7] See CRS Report R41325, Carbon Capture: A Technology Assessment, by Peter Folger.

[8] See CRS Report R42496, Carbon Capture and Sequestration: Research, Development, and Demonstration at the U.S. Department of Energy, by Peter Folger, for further discussion of FutureGen.

[9] Kevin Bliss et al., "A Policy, Legal, and Regulatory Evaluation of the Feasibility of a National Pipeline Infrastructure for the Transport and Storage of Carbon Dioxide," Interstate Oil and Gas Compact Commission, September 10, 2010, Table 3, http://www.sseb.org/ downloads/ pipeline.pdf. By comparison, nearly 500,000 miles of pipeline operate to convey natural gas and hazardous liquids in the United States.

[10] IPCC Special Report, p. 181.

[11] One metric ton of CO_2 equivalent is written as 1 tCO_2; one million metric tons is written as 1 MtCO2; one billion metric tons is written as 1 $GtCO_2$.

[12] IPCC Special Report, p. 31.

[13] These issues are discussed in more detail in CRS Report RL33971, Carbon Dioxide (CO_2) Pipelines for Carbon Sequestration: Emerging Policy Issues, by Paul W. Parfomak, Peter Folger, and Adam Vann, and CRS Report RL34316, Pipelines for Carbon Dioxide (CO_2) Control: Network Needs and Cost Uncertainties, by Paul W. Parfomak and Peter Folger.

[14] IPCC Special Report, p. 31.

[15] Sedimentary basins refer to natural large-scale depressions in the Earth's surface that are filled with sediments and fluids and are therefore potential reservoirs for CO_2 storage.

[16] Data from 2006. See DOE, National Energy Technology Laboratory, Carbon Sequestration Through Enhanced Oil Recovery, (March 2008), at http://www.netl.doe.gov/publications/ factsheets/program/Prog053.pdf.

[17] IPCC Special Report, p. 203.

[18] The Carbon Capture and Sequestration Technologies Program at MIT, Carbon Capture and Sequestration Project Database, In Salah Fact Sheet, http://sequestration.mit.edu/ tools/projects/in_salah.html.

[19] IPCC Special Report, p. 204.

[20] MIT Carbon Capture and Sequestration Project Database, Weyburn Fact Sheet, http://sequestration.mit.edu/tools/ projects/weyburn.html.

[21] DOE Office of Fossil Energy; see http://www.fossil.energy.gov/programs/ sequestration/ geologic/index.html.

[22] IPCC Special Report, p. 223.

[23] See CRS Report R42496, Carbon Capture and Sequestration: Research, Development, and Demonstration at the U.S. Department of Energy, by Peter Folger, for more information on the DOE programs.

[24] See, for example, Mike Soraghan, "Drilling Waste Disposal Risks Another Damaging Okla. Quake, Scientist Warns," Energywire, April 19, 2012, http://www.eenews.net/energywire/ 2012/04/19/archive/1?terms=earthquake.

[25] See Paul Voosen, "Public Outcry Scuttles German Demonstration Plant," Greenwire, December 6, 2011, http://www.eenews.net/Greenwire/2011/12/06/ archive/10?terms= vattenfall.

[26] Carbon Capture and Sequestration Project Database, Sleipner Fact Sheet, http://sequestration. mit.edu/tools/projects/ sleipner.html.

[27] Carbon Capture and Sequestration Project Database, Snohvit Fact Sheet, http://sequestration.mit.edu/tools/projects/ snohvit.html

[28] Carbon Capture and Sequestration Project Database, Gorgon Fact Sheet, http://sequestration. mit.edu/tools/projects/ gorgon.html.

[29] Coal bed and coal seam are interchangeable terms.

[30] IPCC Special Report, p. 217.

[31] U.S. Dept. of Energy, National Energy Technology Laboratory, 2010 Carbon Sequestration Atlas of the United States and Canada, 3rd ed. (November 2010), 160 pages. Hereinafter referred to as the 2010 Carbon Sequestration Atlas, http://www.netl.doe.gov/technologies/ carbon_seq/refshelf/atlasIII/2010atlasIII.pdf; and 2008 Carbon Sequestration Atlas of the United States and Canada, 2nd ed. (November 2008), 140 pages. Hereinafter referred to as the 2008 Carbon Sequestration Atlas. (Available from CRS.) A 2007 Carbon Sequestration Atlas was also published and is available from CRS.

[32] 2008 Carbon Sequestration Atlas, p. 8.

[33] Sean T. Brennan et al., A Probabilistic Assessment Methodology for the Evaluation of Geologic Carbon Dioxide Storage, USGS, Open-File Report 2010-1127, 2010. The USGS has also released at least one report on the geologic framework of specific geologic basins that may help improve estimates of sequestration capacity. See, for example, Jacob A. Covault et al., Geologic Framework for the National Assessment of Carbon Dioxide Storage Resources-Bighorn Basin, Wyoming and Montana, USGS, Open-File Report 2012-1024-A, 2012, http://pubs.usgs.gov/of/2012/ 1024/a/.

[34] 2008 Carbon Sequestration Atlas, p. 23.

[35] 2010 Carbon Sequestration Atlas, p. 139.

[36] Christopher L. Sabine et al., "Current Status and Past Trends of the Global Carbon Cycle," in C. B. Field and M. R. Raupach, eds., The Global Carbon Cycle: Integrating Humans, Climate, and the Natural World (Washington, DC: Island Press, 2004), pp. 17-44.

[37] 2007 IPCC Working Group I Report, pp. 514-515.

[38] CO_2 forms carbonic acid when dissolved in water. Over time, the solid calcium carbonate ($CaCO_3$) on the seafloor will react with (neutralize) much of the carbonic acid that entered the oceans as CO_2 from the atmosphere. See David Archer et al., "Dynamics of Fossil Fuel CO_2 Neutralization by Marine $CaCO_3$," Global Biogeochemical Cycles, vol. 12, no. 2 (June 1998): pp. 259-276.

[39] One study, for example, suggests that the efficiency of the ocean sink has been declining at least since 2000; see Josep G. Canadell et al., "Contributions to Accelerating Atmospheric CO_2 Growth from Economic Activity, Carbon Intensity, and Efficiency of Natural Sinks," Proceedings of the National Academy of Sciences, vol. 104, no. 47 (Nov. 20, 2007), pp. 18866-18870.

[40] For more information on ocean acidification, see CRS Report R40143, Ocean Acidification, by Eugene H. Buck and Peter Folger.

[41] IPCC Special Report, p. 285.

[42] A CO_2 hydrate is a crystalline compound formed at high pressures and low temperatures by trapping CO_2 molecules in a cage of water molecules.

[43] K. Z. House, et al., "Permanent Carbon Dioxide Storage in Deep-Sea Sediments," Proceedings of the National Academy of Sciences, vol. 103, no. 33 (Aug. 15, 2006): pp. 12291-12295.

[44] P. G. Brewer, et al., "Deep Ocean Experiments with Fossil Fuel Carbon Dioxide: Creation and Sensing of a Controlled Plume at 4 km Depth," Journal of Marine Research, vol. 63, no. 1 (2005): p. 9-33.

[45] IPCC Special Report, p. 279.

[46] Ibid., p. 298.

[47] Virginia Gewin, "Ocean Carbon Study to Quit Hawaii," Nature, vol. 417 (June 27, 2002): p. 888.

[48] Jim Giles, "Norway Sinks Ocean Carbon Study," Nature, vol. 419 (Sept. 5, 2002): p. 6.

[49] Serpentine and olivine are silicate oxide minerals—combinations of the silica, oxygen, and magnesium—that react with CO_2 to form magnesium carbonates. Wollastonite, a silica oxide mineral containing calcium, reacts with CO_2 to form calcium carbonate (limestone). Magnesium and calcium carbonates are stable minerals over long time scales.

[50] Calera, a company based in Los Gatos, CA, has developed a process for mineral carbonation that it claims will sequester CO_2 and produce solid carbonate minerals that can be used in the manufacture of building materials. The Calera process is discussed in a CRS congressional distribution (CD) memorandum, available from Peter Folger at 7- 1517.

[51] IPCC Special Report, p. 40.

[52] Flood basalts are vast expanses of solidified lava, commonly containing olivine, that erupted over large regions in several locations around the globe. In addition to the Columbia River Plateau flood basalts, other well-known flood basalts include the Deccan Traps in India and the Siberian Traps in Russia.

[53] 2010 Carbon Sequestration Atlas, p. 30.

INDEX

D

E

F

G

H

I

K

L

M

T

U

V

W